Stability of Complex Carbohydrate Structures
Biofuels, Foods, Vaccines and Shipwrecks

Stability of Complex Carbohydrate Structures
Biofuels, Foods, Vaccines and Shipwrecks

Edited by

Stephen E. Harding
National Centre for Macromolecular Hydrodynamics, University of Nottingham, Sutton Bonington, UK
Email: steve.harding@nottingham.ac.uk

RSC Publishing

Based on the Proceedings of Stability and Degradation of Complex Carbohydrate Structures: Mechanisms and Measurement discussion meeting organised by the RSC Biotechnology and Carbohydrate Groups held in London on 5th September 2011

100683750 6

Special Publication No. 341

ISBN: 978-1-84973-563-6

A catalogue record for this book is available from the British Library

Published by The Royal Society of Chemistry,
Thomas Graham House, Science Park, Milton Road,
Cambridge CB4 0WF, UK

Registered Charity Number 207890

Visit our website at www.rsc.org/books

Printed in the United Kingdom by CPI Group (UK) Ltd, Croydon, CR0 4YY, UK

Preface

Large molecular weight carbohydrate polymers and assemblies are well known for their importance in foodstuffs, paper and wood, but their importance extends far beyond that into the biopharmaceutical, healthcare, oil and printing industries. In many cases their *degradation* is highly favourable – for example in nutrition, paper/agro-waste, recyclisation and their use as potential biofuels. In other instances *retarding degradation* is preferred – in their use as biopharmaceuticals for example, or in preserving old wood structures in archaeology.

A special one-day discussion meeting of the Royal Society of Chemistry in conjunction with the European Polysaccharide Network of Excellence (EPNOE), was therefore recently convened at the Royal Society of Chemistry offices in Burlington House, Piccadilly London. This meeting brought together some of the key representatives from the polysaccharide and glycoconjugate communities to review and discuss in detail the importance of the stability and degradation of these substances, the biochemical mechanisms involved and the latest methods for studying stability and degradation at the molecular level. It was then agreed that such was the importance and interest a book should be produced to provide a permanent record of these discussions.

This book therefore captures the essential parts of that meeting and develops some of the ideas and concepts even further. Professor Felix Franks (Bioupdate Foundation, London) in *Carbohydrates: First cousins of water* considers the structure of polysaccharides and shows how it is important when considering stability issues just how close to water these "polyhydrate" molecules are. Water as an amphipathic molecule, can participate in a wide range of reactions – proton donor/acceptor, oxidation/reduction, and hydrolysis/aggregation – and to some extent carbohydrates and polysaccharides are able to participate in the same types of reactions, although reaction rates in fused or vitreous materials will differ vastly from aqueous solution.

Professor Richard Tester and Chia-Long Lin (Glasgow Caledonian University) then in *Enzymatic stability of starches* consider the structure of starch granules - the composition and architecture – and how this structure provides molecular stability which is essential for energy storage. They examine how via biosynthesis and then exposure to enzymes, starches are stable in aqueous environments. Professor Greg Tucker (University of Nottingham) in his contribution on the *Enzymatic degradation of cell wall polysaccharides* describes the current position in our knowledge of the action and the genetics underlying the action and control of pectinesterase enzymes in fruit – the ripening followed by cell-wall degradation of tomatoes being the classical example. In a related contribution on the stability of cell-wall polysaccharides, Dr. Terri Grassby, Professor Peter Ellis and colleagues (Kings College London) then consider in *Functional components and mechanisms of action of 'dietary fibre' in the upper gastrointestinal tract* how the rate and extent of starch and lipid digestion in the gut can be strongly affected by edible plant polysaccharides, notably the water-soluble viscosity-producing polymers such as legume galactomannans. They consider how studies on almond seeds have shown that the physical state of dietary fibre is crucial, so that intact cell walls post-ingestion can encapsulate lipids and therefore hinder the digestion process.

The stability of polysaccharide and glycoconjugates in biopharmaceutical vaccines is then considered in *Polysaccharide and glycoconjugate vaccines based on bacterial surface glycans* by Dr. Chris Jones (National Institute of Biological Standards, Potters Bar, UK), and how a detailed analysis of biophysical properties is showing how different strategies for glycosylation leads to different stabilities against glycan degradation, and is providing valuable insights into the mechanism by which they degrade. A further important class of glycoconjugates are the mucin glycoproteins, which form an integral part of the mucosal defensive barrier at surfaces throughout the body. As the barrier is closely integrated with innate and immune defensive systems it has a requirement for controlled turnover while retaining defensive functions on a continuous basis. Dr. Tony Corfield (University of Bristol) describes the current understanding of *Mucin turnover* – this is a balance between biosynthesis, oligomerisation and network formation, physical disruption at the mucosal surface and the action of specific enzymes which degrade the glycan chains and the peptide backbone. An important feature of this process is the enzymes which generate and cleave disulphide bridges found within the mucin monomers themselves, and also between monomers to yield dimers and oligomers. Evidence suggests that the formation of mucin fragments is a fundamental part of the turnover process and that these are generated on a tissue- and mucin-specific basis.

Professor Stephen Harding (National Centre for Macromolecular Hydrodynamics, Nottingham, and EPNOE) then considers the state of the art of hydrodynamics in his chapter *Viscometry, analytical ultracentrifugation and light scattering probes* and how we can accurately assess the molecular integrity in solution (molecular weight distribution, conformation and flexibility, aggregation state) of mucins, glycoconjugate vaccines and polysaccharides. Dr. Gordon Morris (University of Huddersfield) and colleagues then in *Stability of pectin-based drug delivery systems* consider the application of a variety of hydrodynamic probes to a study of the stability of pectins extracted from the cell wall of fruits and used in drugs like Fentanyl. Pectins are seen to depolymerise after prolonged storage periods (up to 1 year) and the rate of depolymerisation is considerably greater at elevated temperatures (40°C). This results in a decrease in the strength for low methoxy pectin gels, although this decrease in gel strength was not seen to have a detrimental effect on the release of a model drug. Dr. Christine Wandrey (Ecole Polytechnique Fédérale de Lausanne, Switzerland) and colleagues in *Stability of polysaccharide encapsulation complexes* then consider the importance of hydrophilic nanocarriers as vectors in biomedical and pharmaceutical applications, particularly those involving complexes of chitosan with alginate. They also show how hydrodynamic methods (dynamic light scattering, zeta potential and analytical ultracentrifugation) deliver reliable characteristics with a clear dependence on ionic strength and temperature (as expected). Also strong deformability and stability in different media have been reported as crucial parameters, and this appeared to be the case if either fungal or animal sources of the chitosan prior to linkage with tri-polyphosphate are involved.

The final part of the book considers a series of papers on the stability of cellulose and lignins. Professor Michael Jarvis (University of Glasgow) writing in *Cellulose crystallinity: perspectives from spectroscopy and diffraction* describes how, because of the stability of crystalline cellulose, the degree of crystallinity has a large effect on rates of cellulose degradation in almost every situation where cellulose is degraded, and how X-ray diffraction, solid-state nuclear magnetic resonance, Fourier Transform Infra-red spectroscopy and neutron scattering are giving us powerful insights into this stability and in particular the role of the interplay between water immobilisation and microfibril structure. Professor Callum Hill and Yanjun Xie (Edinburgh Napier University) in their contribution *The water vapour sorption properties of cellulose* then show the importance of moisture content on the stability of cellulose and lignin structures in terms of the parallel exponential kinetics (PEK) model using Kelvin-Voigt based considerations. In the case of a cellulosic fibre subjected to a change in relative humidity, there is a change in the swelling pressure exerted within the cell wall when the atmospheric water vapour pressure is raised.

Professor Claire Halpin (University of Dundee) and colleagues then in *Lignin* consider how this substance is a major barrier to the efficient conversion of wall complex polysaccharides (cellulose and hemicellulose) into the simple sugars that can be used as substrates for liquid biofuel and speciality chemical production. The lignin content and composition can be manipulated at a genetic level by altering the expression of lignin biosynthesis genes and without depending on the gene being targeted - impacting on plant health and fitness, even in field-grown plants. An understanding of these processes is providing valuable information that can be used to devise effective strategies for the future breeding of plant biomass improved for bioenergy applications.

The issue of celluloses and lignin stability form an important part of the final contribution which considers the stability of archaeological wood structures. Nanna Bjerregaard Pedersen (University of Copenhagen) and colleagues in their presentation on *Bacterial degradation of historical wood timbers found in near anoxic waterlogged environments* show how our ever increasing knowledge at the molecular and microbiological level is proving crucial to our understanding of the stability and the assessment of decay profiles of buried ancient shipwrecks. Historical wood timbers can survive for hundreds or thousands of years in anoxic waterlogged environments due to the lack of oxygen supply for the aggressive wood degrading fungi, wood boring insects and wood-boring molluscs. Erosion bacteria cause decay of the secondary cell wall in anoxic environments. They consider how recent research is providing insight into the stability of carbohydrate and lignin in wood structures with ageing and which carbohydrate structures are being preferentially decayed by erosion bacteria.

I hope that this volume in the Royal Society of Chemistry Special Publication Series will provide a useful interdisciplinary insight of value to researchers from across the wide diversity of fields where the stability of carbohydrate and carbohydrate complex structures is important. In bringing these contributions together into one volume I hope this will also lead to further fusion of ideas across the communities working in Plant & Food Sciences, Pharmaceutical Sciences, Obesity research, Biofuels/bioenergy research, Biophysical/ Biochemical and Microbiological analysis and Archaeology.

Finally, I would like to thank the Chemistry-Biology Interface Division and the Carbohydrate and Biotechnology Groups of the Royal Society of Chemistry, together with the European Polysaccharide Network of Excellence, Nestle Ltd. (York, UK) and Glycomix Ltd. (Reading, UK) for their generous support, and Mari Wickerts and Katharina Wänseth of the Göteborgs Stadsmuseum, Sweden for their help with the cover illustration.

Stephen E. Harding
University of Nottingham
October 2012

Contents

CARBOHYDRATES: FIRST COUSINS OF WATER

F. Franks

BioUpdate Foundation, 25 The Fountains, Ballards Lane, London N3 1NL
asdi35@dsl.pipex.com

1 INTRODUCTION

Water has frequently been described as the most eccentric molecule in our ecosphere. Its many anomalous physical properties derive basically from the sp^3 hybridisation of the oxygen orbitals, which gives rise to an almost tetrahedral, quadrupolar bond orientation, in which the oxygen atom is placed at the centre of the water tetrahedron and four charges, two –OH groups (positive) and two lone electron pairs (negative) are directed towards the vertices, as shown in Figure 1. That, in itself, is not a unique molecular feature; for instance SiO_2 and GeO_2 have similar tetrahedral configurations. The unique features of H_2O are twofold: 1) its inability to form or participate in stable covalent bonds. Hydrogen bonding is thus the only method by which the molecule can interact with other molecules, including other water molecules, and 2) because of its quadrupolar nature with an equal number of proton donor and acceptor sites, H_2O is thus amphipathic, and as such, it can participate in a wide range of reactions: proton donor/acceptor, oxidation/reduction, and hydrolysis/aggregation. To some extent, carbohydrates are able to participate in the same types of reactions, although the rates in fused (vitreous) carbohydrates will differ vastly from those in liquid aqueous solutions.

The description of molecules composed of C, H and O as 'carbohydrates' provides a clue for some of their properties that are of particular significance to their industrial and medical applications. Another generic description is polyhydroxy compounds (PHC), which provides a further clue – water sensitivity/miscibility – to their usefulness. Added to this are their diverse functions in life processes, providing structures, energy storage, metabolic intermediates, recognition and defence (immune) systems and much else. The generic formula of many simple PHC compounds, when written as $C_n(H_2O)_m$, also suggests a close relationship between water and PHC molecules. Indeed, any material in our ecosphere that is composed mainly of C, H and O is either miscible with water, or at least sensitive to water, so that dry (anhydrous) organic materials do not exist in nature. Water has been described as the 'ubiquitous plasticizer' for all organic matter.

The actual relationship between water and PHCs at the molecular level lies in their almost identical oxygen orbital hybridization: -OH groups of the organic molecules structurally and energetically resemble those of water, i.e. in their interactions they are limited to hydrogen bonding and their bond orientations are also in the form of a tetrahedron. It indicates that PHC molecules in their crystalline state closely resemble ice

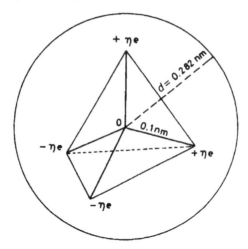

Figure 1 *The four-point-charge model of the water molecule. The oxygen atom is placed at the centre of a regular tetrahedron, the vertices of which are occupied by two positive (hydrogen atoms) and two negative (lone electron pairs) charges, the O-H distance being 0.1 nm. The distance of closest approach of two molecules (van der Waals diameter) is 0.282 nm*

and are held together (or apart) by weak interactions and, like ice, they can form chains, rings and infinite three dimensional networks. In their crystalline states they can also interact with water in the form of stoichiometric hydrates, e.g. glucose.H_2O, α,α-trehalose.$2H_2O$, β,β-trehalose.$4H_2O$, raffinose.$5H_2O$. Like inorganic hydrates, for example Na_2CO_3.$10H_2O$, they should be able to be dehydrated to lower hydration states and finally to anhydrous crystals, but unlike their inorganic counterparts, most of these hypothetical solid/solid transitions take place very slowly and proceed via long-lived non-crystalline (amorphous) intermediate states; characteristic transition temperatures cannot therefore be easily observed. Their phase transitions usually take place over very extended periods (months or years), and very few such systems have actually been studied directly in 'real time'.[1,2] Therefore, the dehydrations appear to be irreversible in real time, so that the mixtures may consist of the higher and lower hydration states and the released water. The mixture will thus remain thermodynamically unstable and maintain an apparent stability due to exceedingly slow molecular motions, in the order of mm/century, until a spontaneous devitrification to the more stable, crystalline state, e.g. during annealing, occurs and is accompanied by more or less severe changes in their mechanical properties, such as tensile or torsional strength. Examples of such spontaneous devitrifications readily occur in metal alloys, in the crystallisation of cholesterol from its highly supersaturated solution in the gall bladder, or in the crystallisation of lactose in ice cream. The results may be of different orders of severity, ranging from loss of life to physical pain and minor inconvenience. Thus, several of the first ever jet-engined commercial airliners, the *de Havilland Comet*, disappeared during long-haul flights, when the fuselage disintegrated, resulting from metal stress and subsequent devitrification.[3]   The nucleation and biocrystallisation of cholesterol in the gall bladder results from a chemically minor error in the biosynthesis of bile salt molecules, whose natural function is the prevention of cholesterol precipitation from its supersaturated solution.[4] The possibility of random devitrification of lactose in some cartons of ice cream, even during frozen storage, is still a subject of research.[5]

Other possible sources of PHC dehydration problems include the decomposition of a previously unknown PHC hydrate, especially in pharmaceutical formulations, where such release of water during processing, e.g. mannitol. $H_2O \rightarrow$ mannitol + H_2O, can result in major stability losses in the drug product.[6] With our present sketchy knowledge of the close relationship between water and the hydrogen bond topology of PHCs, it is likely that more such presently unknown hydrates remain to be discovered.

It is also likely that future experimental data on crystalline PHC dehydration will remain to be of uncertain value, because of the extended periods required for these reactions to go to completion.

2 THE CONCEPT OF PHC HYDRATION AND ITS MEASUREMENT

To study binary mixtures of molecules which differ substantially in size and shape but closely resemble each other in their interactions (hydrogen bonds), certain simplifications are necessary. In the past, the concept of 'bound water' was developed to provide explanations to the above problems. In its simplest form it consisted of covering the exposed area of the PHC molecule with notionally spherical water molecules and counting the minimum number that would completely cover the PHC. Eventually experimental techniques, such as neutron scattering became available to verify (or discount) such simple approaches. They proved useful to establish the geometry of ion hydration and the assignment of credible hydration numbers, especially for monatomic ions.[7]

However, even with molecules such as mono-, di-, or oligosaccharides in their aqueous mixtures, and leaving aside polymeric PHCs, such approaches can hardly be considered as satisfactory. In addition, the PHC in solution often exists as mixtures of two or more anomeric states. Finally, the sciences of physics and physical chemistry do not recognise the existence of different species of H_2O, bound or free. In order to provide a measure of credibility for such a distinction, certain criteria must be established as to how 'bound' can be defined for the water-PHC system. At its simplest, these criteria are based on structure, energetics and dynamics:

(i) **Structure**: measured in terms of distances, orientations and Euler angles, by X-ray or neutron diffraction and circular dichroism

(ii) **Energetics:** the existence of special, probably favourable interaction energies of certain hydrogen bonds in specific locations, measured (with difficulty) by thermodynamic P-V-T properties, calorimetry and volumetry; also in principle by infrared or Raman spectroscopy

(iii) **Dynamics:** exchange rates of water molecules considered to be bound, their rotational and/or translational diffusion rates, measured by relaxation techniques, such as dielectric relaxation or nuclear magnetic resonance (NMR). A critical evaluation of the above, and other techniques, as applied to studies of PHC-water systems, is given in ref. 8.

3 HYDROGEN-BOND NETWORKS IN PHC-WATER SYSTEMS

In PHC-water mixtures, the following hydrogen bonds may be considered, where OW and OC refer to oxygen atoms of water and a PHC molecule respectively, and the colon ":" refers to a proton acceptor site:

WO-H → :OC   CO-H → :OW   WO-H → :OW   CO-H → :OC   etc.

The study of hydrogen bonding in its various aspects must centre on the behaviour of the exchangeable proton and details of the bond geometry. Unfortunately, spectroscopy has for long been obsessed with the desirability of narrow signals, and it was for many years a habit to exchange $-O^1H$ protons by $-O^2H$, thereby simplifying the spectra, but also removing the spectral features that provide the information about hydrogen bonds. Bearing in mind that bond lengths and angles in water are almost identical with those in PHCs, direct studies of hydration must in most cases be subject to major experimental problems.

In attempts to study interactions between sugar –OH groups and water, Symons and co-workers first observed that over narrow pH ranges and at low temperatures, high resolution ^1H NMR signals due to sugar –OH groups could be resolved at 1 – 3 ppm downfield from the H_2O signal, measured at 100 MHz.[9,10] For the identification of a *structural* influence on hydrogen bonding in binary sugar-water mixtures, compare hypothetical hydration systems in hexoses and pentoses in their α- with β-anomeric forms, respectively. Anomers differ only in the configuration of the –OH groups on C(1) (axial *versus* equatorial). At ordinary temperatures the proton transfer WO-H → :OC(1) is fast, and any hydration difference between the α and β anomers could for a long time not be directly measured. However, as shown in Figure 2, by inserting an α- or β-hexopyranose molecule into the notional tetrahedral 'structure' of liquid water, but without causing strain in the hydrogen bonded network, it is apparent that a significantly better hydration fit is provided for the β-C(1)O-H interactions with water, than can be produced with the α-anomer.

In order to establish the nature and possible effects of any such differences, an extensive series of experimental studies was initiated by a group of scientists at the Unilever Colworth Research Laboratories during the 1970s. The methodologies and results are summarised in ref. 8. The techniques included calorimetry, PVT properties, dielectric relaxation (frequency and time domain) and NMR relaxation (^{13}C, ^1H, ^2H, ^{17}O) of several mono- and oligosaccharides in aqueous solution. It was a happy coincidence that at the same time nucleation and crystallisation processes of water at subzero temperatures were under intensive study[11], and we developed practical methods for achieving the undercooling of liquid water down to ca. -40°C, without the necessity of adding cryoprotective agents.[12] This made possible the slowing down and the resolution of individual steps in coupled multistep processes.

The technique proved to be particularly useful for the separation and analysis of individual steps in complex biochemical processes, e.g. bioluminescence.[13] We applied the techniques of deep undercooling to the study of proton residence times and exchange rates for several sugar-water systems and obtained time resolved ^1H spectra from which proton residence times and activation energies of the proton transfer C(1)O-H → :OW could be calculated for glucose and a number of related sugars,[14] in particular the corresponding pentose anomers of xylose.[15] It had already been observed that the chemical shifts for xylose are almost identical with those of glucose (although at different pH values), indicating that the CH_2OH group on C(5) of xylose affects neither the acidity nor the solvation of the anomeric –OH. This conclusion is further confirmed by the very similar

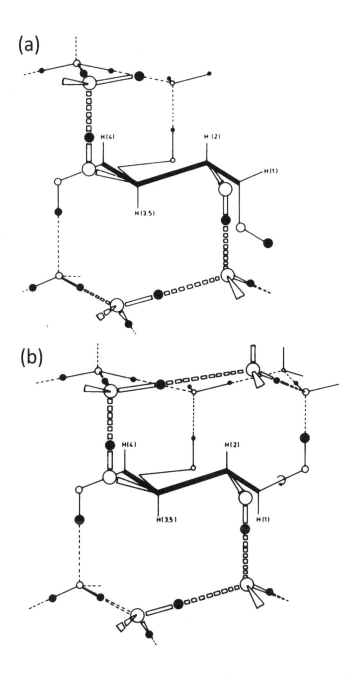

Figure 2 *D-glucose hydrogen bonded into a hypothetical water structure, according to the specific hydration model. Water molecules below and above the plane of the sugar ring are shown: (a) α-anomer and (b) β-anomer. The pyranose ring is indicated by the bold line. Oxygen and hydrogen atoms are represented by open and filled-in circles; covalent and hydrogen bonds by solid and broken lines. The hydroxymethyl protons (H(6)) are omitted for the sake of clarity. Reproduced, with alterations, from ref. 18*

anomeric equilibrium compositions of the two sugars in aqueous solution. On the other hand, the positions of the ring –OH groups markedly affect the –OH shifts. This is seen by comparing the pentoses xylose and ribose (in ribose the –O(3)H is axial) with the hexoses glucose and mannose (*axial* –O(2)H), confirming previous suggestions that the hydration of sugars is subject to a high degree of stereospecificity.[14,16]

Dissimilarities in the thermodynamic solution properties, the free energy ΔG and its T and P derivatives, of ribose and xylose also become apparent from a comparison of their osmotic coefficients (the second virial coefficient in the osmotic pressure equation). Whereas aqueous solutions of ribose are close to ideal, xylose exhibits marked positive deviations from ideal behaviour, indicating a net repulsion between xylose molecules.[17]

The above discussion of pairs of single, but related sugar molecules in a hypothetical aqueous environment may seem laborious and convoluted, but it points to distinct differences in the interactions between water and chemically very similar carbohydrate molecules. These features have been classified as examples of 'specific hydration', i.e. they are exclusively observed in *aqueous* solutions and are related to differences in the stereochemical details of hydrogen bonds between water and specific -OH sites of PHC molecules.[18]

4 PRACTICAL ASPECTS OF THE 'FIRST COUSIN' RELATIONSHIP

The isolated molecule and hypothetical water structure approaches are of limited practical value. Rather, the bulk properties of highly concentrated, supersaturated mixtures, i.e. the material science aspects, need to be considered. Tests must be applied whether the special relationship between PHC molecules and tetrahedrally 'structured' water molecules yields useful information about bulk properties such as undercooling, supersaturation, vitrification, crystallisation, viscous flow, elastic moduli, structural and chemical stability, etc.

Fortunately, here again there exists an excellent series of publications that probe such relationships.[19-22]   Basically, the experiments describe the rotational and translational diffusion of water relative to those of different parts of the PHC molecules, as the concentration increases to beyond the saturation value. The techniques encompassed ^{13}C NMR for the PHCs and –OD, -CH$_2$OD and D$_2$O, a so-called "hydration phase", by ^{2}H NMR over a temperature range (220-350 K) and concentrations (10-70% w/w) for sucrose, α,α-trehalose, maltose and   α-D-methyl glucoside. The following three questions were addressed, namely: 1. How do PHCs affect the dynamics of water?   2. How do the dynamics of PHC molecules depend on the concentration?   3. What conditions determine the transition (glass transition T$_g$) into the vitreous state?

In summary, some of the results of the Girlich thesis can be discussed in terms of three distinct concentration domains (Figure 3).

4.1 The case of the concentration, c ≤ 30%

PHC molecules perform uncorrelated motions, and the solution viscosity exhibits a minimal concentration dependence. The dynamics of water molecules are governed by cooperative fluctuations within the hydrogen-bonded water network. In undercooled solutions the network increasingly inhibits the orientations and the rotational correlation time (τ) increases steeply with decreasing temperature. Hydrostatic pressure distorts the network and, in the metastable phase, it inhibits the generation of ordered domains with linear H-bonds; molecular motions thus remain rapid. Dissolved PHC molecules, because of their incompatible –OH orientations, produce a similar effect. This facilitates the undercooling of PHC solutions far into the metastable region, resulting in eventual vitrification, rather than crystallisation.

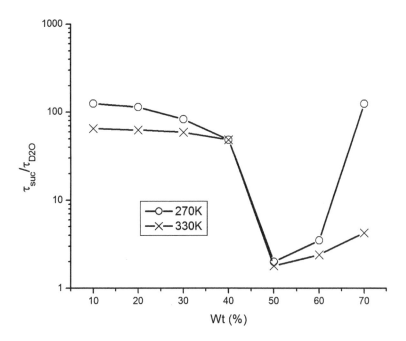

Figure 3 *Ratio of NMR rotational relaxation time (τ) of sucrose to that of water as a function of sucrose concentration, showing the phenomenon of maximum dynamic coupling between sucrose and water at 50% w/w (4 mol%), which disappears as the glass transition is approached. Redrawn from ref. 22*

4.2 The case of the concentration, c > 40%

Hydrated PHC molecules form aggregates which develop into a network. The resulting change in the short-range structure leads to a distribution of correlation times. The water molecules are integrated into the network, and their rotational and translational motions are affected. This is strikingly demonstrated in Figure 3, which shows the ratio of sucrose:water τ values as a function of sucrose concentration. An increasing coupling of diffusive motions (microheterogeneity) between sugar and water is evident in a complex mixture; it reaches its maximum value at c ~ 50%. At higher concentrations, PHC molecules form networks by direct H-bond links, with the expelled water molecules collecting in microscopic droplets; their mobility is facilitated in this quasi-heterogeneous system. At this stage PHC and water motions lose their coupled motions, with a sharp rise in the τ ratio.

4.3 The case of the concentration, c > 70%

A three-dimensional PHC network is formed. With decreasing temperature, the network forms a macroscopic gel, and the PHC molecules are unable to adopt the crystal configurations in real time. The metastable mixture forms a glass. Water molecules are trapped in the free volume within the glass; their translational diffusion is (almost) blocked, i.e. they become osmotically inactive, although their rotational freedom persists even down to 120 K.

Unfortunately the authors, like many other workers, adopted the w/w concentration scale, which simplifies calculations but hides more information than it reveals. A more appropriate scale for comparisons of different PHCs is the mol percent concentration. Thus a 50% w/w solution of sucrose (and other disaccharides) corresponds to 5 mol% sucrose and 95 mol% water; hardly a concentrated solution, although its viscosity might suggest the opposite.

None of the above dramatic dynamic effects are revealed in the macroscopic thermodynamic properties of PHC-water mixtures, nor can they be studied by X-ray diffraction, because of the short range of the microheterogeneities. In principle, neutron scattering might provide more information, but at the time of writing, few reports have yet found their way into the public domain, except for a recent neutron scattering study of sorbitol-water mixtures.[2] However, the comprehensive studies by Girlich (1991) of PHC solutions involving multinuclear NMR relaxation measurements over extended pressure-temperature-composition phase domains, are of a quality rarely found in the scientific literature; they have revealed much detailed information of relevance to ecology, climate physics and several branches of medicine and pharmaceutics that depend on the dehydration of sensitive materials.

5 CONCLUSIONS

Over the past 50 years, new insights have been developed from the first ever detailed studies of PHC-water mixtures. Thus, it is now accepted that the reluctance of PHCs, with the exception of mannitol, to crystallize from aqueous environments is caused by the marked similarities in the hydrogen bonding properties of the two substances. This realization took a long time to gain acceptance by the carbohydrate community. During a visit to the Materials Science Department at the University of Cambridge I mentioned the term 'water soluble glasses' and received looks of utter disbelief. It was only when I said: *'sugar candy!'* that some recognition showed in the faces of the audience. This recognition eventually led to the birth of a new branch of Materials Science: aqueous amorphous solids (glasses) of materials which, unlike crystals, possess no long range molecular order, but which, although thermodynamically metastable, can over limited periods (shelf lives) be treated as stable solids as regards their bulk properties. They derive this apparent stability not from a favourable free energy of mixing, but from slow kinetics, where the transitions between states are subject to very high activation barriers.[23]

Before setting out on the long path to the elucidation of PHC hydration, Alan Suggett had already concluded that carbohydrates were the *'Cinderellas of Biophysical Chemistry'*.[24] At the time, the limelight was all on DNA, proteins and lipids, as were also the Nobel Prizes.

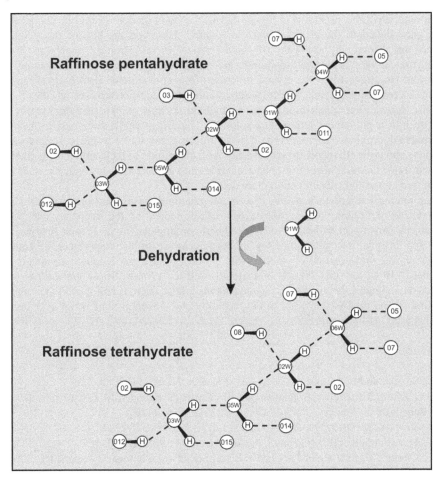

Figure 4 *The dehydration of raffinose.5H$_2$O to raffinose.4H$_2$O through the elimination of W(1). Changes in the observed crystal structure are almost completely limited to the positions and spacings of the hydrogen bonds between water and sugar molecules*

Several years ago we began to study the dehydration reaction

$$\text{raffinose.5H}_2\text{O (s)} \rightarrow \text{raffinose.4H}_2\text{O (s)} + \text{H}_2\text{O} \qquad (1)$$

This necessitated the ability to remove specific water molecules from the crystalline pentahydrate, the isolation of single tetrahydrate crystals and their characterization by single crystal X-ray diffraction, followed by a comparison with the known pentahydrate crystal structure.[25]

The crystal structure of raffinose tetrahydrate has now been studied by single-crystal X-ray diffraction at 93K.[26] The molecular structure of the raffinose molecule as well as the hydrogen-bonding scheme in the tetrahydrate crystal closely resemble those observed in pentahydrate. Upon dehydration, the positions of water 1 and 4 in pentahydrate disappear, and instead, a new position of a water molecule, W(6), is observed in the structure of raffinose tetrahydrate, as shown in Figure 4. This results in the loss of infinite chains along the *a* axis, linked to the chains along the *b* and *c* axes, and the loss of many weaker interactions. Moreover, the role of water molecules as stronger acceptors in the pentahydrate crystal is changed to stronger hydrogen bond donors in the tetrahydrate. The results, so far, suggest that the crystal structure of the conventional pentahydrate tends to become increasingly unstable on partial dehydration. The question remains whether further stepwise dehydration might produce lower hydrate crystals, or whether the system will collapse into a permanent amorphous solution. The above studies emphasize the many experimental problems encountered in what might seem to be a fairly simple set of questions about the 'first cousin' relationship. The magnitude of these problems is likely to increase, once the relationship of polysaccharides and glycoconjugates with water *in vitro* and *in vivo* comes to be examined in more detail.

The practical importance of the PHC-water relationships in food process technology was realized more than three decades ago; it can be judged by the frequency with which the subject turns up in national and international conferences.[23,27-29] It took several more years for the pharmaceutical industry to become aware of the importance of aqueous glasses in the formulation of shelf stable products, such as biopharmaceuticals and vaccines.[30] However, this branch of materials science is now firmly embedded in the various technologies that are based on aqueous materials. There is still much to learn about the 'first cousin relationship', but in the meantime, the advantages of water over organic solvents need hardly be stressed.

References

1   T. Suzuki, and F. Franks, *J. Chem. Soc. Faraday Trans.,* 1993, **89,** 3283.
2   S.G. Chou, A.K. Soper, S. Khodadadi, J.E. Curtis, S. Krueger, M.T. Cicerone, A.N. Fitch and E.Y. Shalaev, *J. Phys. Chem. B,* 2012, **116,** 4439.
3   S. Stewart, *Air Disasters.* Arrow Books, Ltd (U.K.) 1986 & 1989.
4   W.H. Admirand and D.M. Small, *J. Clin. Investig.,* 1968, *47,* 1043.
5   C. Clarke, *The Science of Ice Cream*, Royal Society of Chemistry, Cambridge, 2004, p.135.
6   L. Yu, N. Milton, E.G. Groleau, D.S. Mishra and R.E. Vansickle, *J. Pharm. Sci.,* 1999, **88,** 196.
7   J.E. Enderby and G.W. Neilson, in *Water – A Comprehensive Treatise*, Vol. 6, ed. F. Franks, Plenum Press, New York, 1979, ch.1.
8   F. Franks and J.R. Grigera (1990) *Water Sci. Rev.,* 1990, **5,** 187.
9   J.M. Harvey, M.C.R. Symons, and R.J. Naftalin, *Nature*, 1976, **261,** 435.
10   J.M. Harvey and M.C.R. Symons, *J. Solution Chem.,* 1978, **7,** 571.
11   C.A. Angell, in *Water – A Comprehensive Treatise*, Vol. 7, ed. F. Franks, Plenum Press, New York, 1982, ch.1.
12   R.M. Michelmore and F. Franks, *Cryobiology,* 1982, **19,** 163.
13   P. Douzou, C. Balny and F. Franks, *Biochim. Biophys. Acta,* 1978, **540,** 346.
14   S. Bociek and F. Franks, *J. Chem. Soc. Faraday Trans.* 1978, **75,** 262.
15   E. Lai and F. Franks, *CryoLett.,* 1979, **1,** 20.
16   F. Franks, D.S. Reid and A. Suggett, *J. Solution Chem.,* 1973, **2,** 99.
17   H. Uedaira and H. Uedaira, 1969, *Bull. Chem. Soc. Japan* **42,** 2137.
18   A. Suggett, *J. Solution Chem.* 1976, **5,** 33.
19   D. Girlich, *Molecular dynamics of aqueous sugar solutions studied by NMR (in German). Ph.D. thesis*, University of Regensburg, 1991.

20 D. Girlich, and H.-D. Lüdemann, *Z. Naturforsch.,* 1993, **48c,** 407.

21 D. Girlich, H.-D. Lüdemann, C. Buttersack and K. Buchholz, *Z. Naturforsch.* 1994, **49c,** 258.

22 D. Girlich and H.-D. Lüdemann, *Z. Naturforsch.,* 1994, **49c,** 250.

23 H. Levine (ed.) *Amorphous Food and Pharmaceutical Systems,* Royal Society of Chemistry, Cambridge, 2002.

24 A. Suggett, in *Water – A Comprehensive Treatise*, Vol. 4, ed. F. Franks, Plenum Press, New York, 1975, ch. 6.

25 G.A. Jeffrey and D-B. Huang, *Carbohyd. Res.* 1990, **206,** 173.

26 C. Viriyarattanasak, M. Shiro, S. Munekawa, F. Franks, S. Ikeda and K. Kajiwara, *Acta Cryst. C,* 2012 (in press)

27 J.M.V. Blanshard and P.J. Lillford (eds.) *The Glassy State in Foods,* Nottingham University Press, 1993.

28 L. Slade and H. Levine, in ref. 23, p.139.

29 H. Levine and L. Slade, *Water Sci. Rev.*, 1998, **3,** 79.

30 F. Franks and T. Auffret, *Freeze-drying of Pharmaceuticals and Biopharmaceuticals – Principles and Practice*, Royal Society of Chemistry, Cambridge, 2007.

2

ENZYMATIC STABILITY OF STARCHES

C.-L. Lin and R. F. Tester*

Department of Biological and Biomedical Sciences, Glasgow Caledonian University, Glasgow G4 0BA, Scotland, UK
*R.F.Tester@gcu.ac.uk

1 INTRODUCTION

Starch is probably the most important energy reserve in higher plants. It is synthesised in different organelles for 'transitory' as well as the more obvious longer term energy depot roles in seeds, grains, piths, tubers and roots. Depending on which organelle type the starch is synthesised within, it has a characteristic form with different functionality. In chloroplasts the transitory starch (in the form of granules) is synthesised during the day and degraded at night, supporting the growth of the plants or providing energy whenever it is needed. In contrast, the storage/reserve (depot) starch is produced in the amyloplasts of storage tissues (again as granules) during the development of the plants. It is hydrolysed (ultimately to glucose) and utilised as a carbon source during sprouting, germination or ripening of the crop (usually translocated as sucrose). The storage starch can thus be found widely in plant tissues that require a significant energy sink. This type of starch is not only vital for the survival for higher plants survival but also for humans as a food supply. The storage starch (native or modified) has been widely used in the food industry as a texture modifier, stabiliser and thickener and in non-food industries for abrasion resistance improvement in textile manufacturing, for strength and printability enhancement during paper production and as excipient forms in pharmaceutical products. In addition, with a nutritional focus, the hydrolysis resistance of starch to intestinal enzymes has been optimised to provide enhanced nutritional benefits of foodstuffs.

Ungelatinised starches either within or extracted from plant tissues have been studied extensively in terms of their digestibility. Similarly, processed starches have been studied to a large degree. The term, *resistant starch* (RS), has been widely used to describe the form of starchy material which is indigestible in the human small intestine but fermentable in the colon. It can have similar functionality to dietary fibre. Depending on the structural characteristics of resistant starch, it can be defined as: *Type I RS* where the granules are inaccessible by digestive enzymes due to the protection of cell walls and tissue integrity; *Type II RS* which reflects the morphology and crystallinity of different starch granules; *Type III (retrograded) RS* where the granules are subject to processing-induced gelatinisation then re-crystallisation; and *Type IV RS* in which the structural integrity is reinforced chemically (or potentially physically).

The structure of starch granules - the composition and architecture – provides molecular stability which is essential for energy storage. The granules optimise energy deposition in a format which can be mobilised rapidly. This is the key theme of this chapter.

2 BIOSYNTHESIS

The storage starch of plants, unlike the transitory starch, is of high commercial importance as mentioned above and has substantial applications in animal and human diets. Its biosynthesis is a complex process and occurs as granules of approximately 5-80 µm diameter in the amyloplasts of storage tissues. After photosynthesis, sucrose - as the key carbon source - is transported to the storage tissues from the leaves/stems (via the phloem) of plants. There it is converted in the cytosol into adenosine diphosphate-glucose (ADP-glucose) by the action of sucrose synthase, uridine diphosphate-glucose pyrophosphorylase and ADP-glucose pyrophosphorylase (AGPase). The ADP-glucose is then transported directly into developing amyloplasts for starch synthesis in the case of cereal endosperm cells although hexose phosphate may also be transferred (as for other storage tissues)[1,2]. In the storage tissues of potatoes, tapioca and legumes, however, ADP-glucose is synthesised preferentially by amyloplastidial AGPase for which glucose 1-phosphate (G1P) is used as the glucose source derived from the cytosol[1,2]. This enters the amyloplasts from the hexose phosphate/inorganic phosphate (amyloplast membrane) translocator into the amyloplast stroma via the transient form of phosphorylated glucose (glucose 6-phosphate; G6P). After (amyloplast) AGPase activity, the glucosyl unit of ADP-glucose is used as the primer for elongation of α-glucan chains by various starch synthases (SS), which can be simplified as follows:

$$\text{ADP-glucose} + \alpha\text{-glucan} \xrightarrow{\text{starch synthases}} \alpha(1\text{-}4)\text{-glucosyl glucan} + \text{ADP} \qquad (1)$$

The α-glucan chains may be processed further by branching enzymes (BE):

$$\alpha\text{-glucan} \xrightarrow{\text{branching enzymes}} \alpha(1\text{-}4/1\text{-}6)\text{-glucan} \qquad (2)$$

The starch synthases involved in elongating α-glucan chains, depending on their spatial relationships to the starch granule, can be categorised into two classes: soluble starch synthase (SSS) and granule-bound starch synthase (GBSS). The SSS is believed to be predominantly responsible for the synthesis of amylopectin molecules and the GBSS for the synthesis of amylose molecules. The soluble starch synthases can be subdivided further into classes SSSI-SSSIV according to their substrate-specific preferences. Studies on the catalytic property of SSSI suggest that its primary functionality is to incorporate the glucosyl moiety of ADP-glucose into an α-glucan chain with a degree of polymerisation (DP) < 10 although its activity is hindered as the chain length of substrate reaches about DP14 [3,4]. In contrast, SSSII shows lower affinity to the substrate with a heavily branched structure than does SSSI[5,6]. Amylopectin molecules isolated from SSSII deficient mutants consist of lower amounts of the internal chains than do wild types[7,8]. As a consequence, it has been suggested that SSSI is responsible for the elongation of the external chains of amylopectin molecules[4] whereas SSSII has been suggested as responsible for the synthesis of internal chains[7,8]. Compared to the effects of SSSI and SSSII, starch synthesised with suppressed activity of SSSIII shows no alteration in the molecular structure of amylopectin. However, the morphology of granules where deep cracks from the hilum of simple granules and the protrusion of subgranules from the granular surface (similar to the morphology of compound granules) can be observed[9,10]. In comparison with the effects of SSSIII, defects in SSSIV activity lead to drastic alterations to granular morphology in terms of the size but with little changes to the structure of amylopectin[11]. Therefore, SSSIV and SSSIII are thought to be related directly to the establishment of the initial granular structure for starch deposition and the synthesis of starch granule with normal morphology, respectively.

In addition to the classes of soluble starch synthases, GBSS has been shown to be involved in the elongation of α-glucan chains. Unlike the soluble starch synthases, GBSS resides on the developing starch granule rather than in the amyloplast stroma. Starch granules extracted from GBSSI (*waxy*) gene defective mutants are characterised by a substantially lower amylose content than those of the non-mutants[12-14]. Hence, the synthesis of amylose molecules is considered to arise primarily from the activity of GBSSI. Of interest, in addition to there being a lack of amylose, starches formed from the GBSSI defective organism *Chlamydomonas reinhardtii* show a structural alteration in the amylopectin molecules with less internal chains[15,16]. Due to the benefit of the high activity of *Chlamydomonas* GBSSI, the purified starches from unmutated strains can be synthesised *in vitro* by this GBSSI in the presence of primer (ADP-glucose). The resulting starches are characterised by an increased amylose content plus a high percentage of B-type polymorphic structures[17]. Although the GBSSI of higher plants shows 10-50 times lower catalytic activity than that of *Chlamydomonas* strains, the synthases such as in potato and pea starches retain their synthetic properties to elongate amylopectin or amylose *in vitro*. The elongation of the latter is stimulated by the presence of maltosaccharides[18]. This may reflect the fact that the GBSSI has higher affinity for maltosaccharides as the precursor than does the other GBSS isoform (GBSSII). Additionally the GBSSII is expressed in the early development of storage tissue and thus is probably embedded in more confined environment which leads to the *in vitro* elongation by GBSSII being hindered. In addition to the difference in substrate affinity, GBSSI transfers preferentially more than one glucosyl unit onto the precursor before dissociation, whereas a step-wise mechanism is apparent for GBSSII[19].

The formation of α(1-6) glucosidic linkages during starch synthesis arises after the BE (branching enzymes) activity by cleaving an α(1-4) linkage of a primer. This yields a maltosaccharide which is transferred onto an *O*-6 of either the original chain or another α-glucan chain. Unlike the SSSI-IV and GBSSI-II, BE is considered to be localised at the interface between the stroma and the developing starch granule[20]. Two major isoforms of branching enzyme are reported to have distinctively different specific activities on the primer. The BEI has higher activity (~40 times) towards branching linear primers than the pre-branched form, with the opposite profile for BEII (twice the rate of linear primer)[21]. Additionally, the BEII catalyses branched primers at 2.5 - to 6 - fold the rate of BEI and forms shorter side chains than for BEI. It has been proposed that BEI plays a role in the production of slightly branched α-glucans as the substrates for other starch synthases and may also participate in branching the more internal focused chains of amylopectin. The BEII is involved more in the synthesis of branch points on the side chains of amylopectin[21].

With regards the starch debranching enzyme (isoamylase), in addition to SSSI-IV, GBSSI-II and BEI-II, it has been reported to be associated with phytoglycogen formation where it is accumulated in mutants with isoamylase deficiency[22-27]. Based on this observation, it has been proposed that for highly branched precursors, due to the activity of branching enzymes, these need to be trimmed (by the debranching enzyme). This is in order to be a suitable substrate for subsequent catalysis by synthases[22]. Another interpretation of phytoglycogen accumulation is that the dynamic balance between substrates and starch synthases are affected by the reduced or absent activity of isoamylase. The alteration leads to a diversion of the limited amount of other starch synthases to synthesise phytoglycogen[28,29].

Whilst the properties of individual starch synthases have been elucidated to a great extent, the diversified combination of synthases from different species complicates the *in vivo* process of starch synthesis. In addition, the expression of each synthase at different stages of development, their activity on the available composition of substrates/primers and the interplay between each enzyme complicate further the starch synthesis process (and contribute various molecular structures to starches from different species).

As mentioned above, the activity of SSSI in some plants such as potato tubers is lower than that of SSSII, SSSIII and GBSSI[30]. Additionally the expression level of BEII is lower than that of BEI[31]. The combination of lower activities for SSSI and BEII contribute possibly to the amylopectin structure of potato α-glucans with longer unit chain lengths and fewer branch points compared to the molecular structure of amylopectin from cereal starches. Interestingly, potato starch from a double mutant with SSSII and SSSIII deficiency shows holes through the centre of granules; but this abnormal morphology is not found in either of the single mutants[10,32,33]. In the case of the suppression of BEII (amylose-extender mutant maize), the starch is characterised by increased amylose content and amylopectin with longer chain lengths and less branch points.

3 STARCH COMPOSITION

Starch comprises D-glucopyranose polymers representing more than 98% of its dry weight, linked in the polymeric form via α-glycosidic bonds. Two major types of α-glucans, amylose and amylopectin, are synthesised within starch granules with distinctive molecular structures. Amylose is composed of D-glucopyranose residues linked via α(1-4) bonds forming an essentially linear structure. In contrast, amylopectin is a highly branched polymer where ~5% of its α-glycosidic linkages are synthesised as α(1-6) bonds (the branch points) along with the bulk linear α(1-4) linkages.

Starch consists typically of more amylopectin than amylose. The amylose content of maize starches is ~30 ± 4% [34], ~29 ± 1% for wheat starches[35], ~20 ± 4% for tapioca starches[36,37], and ~28 ± 3% for potato starches[38,39]. In addition to the effect of species, the amylose content of starch can vary wildly via forms of mutation. With the waxy enzyme gene deficiency, <3% of amylose content has been reported for maize[34], wheat[40], tapioca[41] and potato starches[42]. Starches with high amylose contents of 42-68% have been obtained for maize endosperms with BE deficiency[34]; >31% amylose for wheat grains with SSSII deficiency with low activities for SSSI and BE[43]; ~30% amylose for tapioca roots with low activity of isoamylase[44] and 27-78% of amylose for potato tubers with BE deficiency[45,46].

The amylose content of starch is also affected by the growing condition of the crop and the response to the environmental effect differs amongst species. As the growth temperature increases, the amylose content of maize starch decreases[47,48], but the opposite effect occurs for wheat[49,50] and potato starches[51-53]. A small environmental effect has been reported for tapioca starch[54]. The agronomic factors also can affect the amylose content of starch[50,55,56]. However, the species and cultivar differences remain the primary factors to affect the amylose content of starch[57].

Lipids exist in cereal starch granules, along with proteins and minerals (particularly phosphorus) as the minor compounds. The lipid content of starch is typically <1% and varies depending on the species of starch (cereal vs root/tuber) and the amylose content. Tapioca and potato starches contain little lipid; generally less than 0.1% [58]. In contrast, the lipid content of cereal starches have been reported to be in the range 0.61-0.77% for maize[34] and 0.77-1.17% for wheat starches[59]. As the amylose content increases, the amount of lipid also increases from <0.14% (waxy maize starches) to 1.03% (an average; high amylose maize starches)[34]. In terms of the lipid composition, it differs wildly between different cereal starches. The lipids existing in *Triticeae* starches are almost only lysophospholipids, while both lysophospholipids and free fatty acids are present for other cereal starches[59]. The origin of true/interior starch lipids associated with the α-glucans remains obscure. However, it has been hypothesised that the lipids are possibly the metabolites of diacylphospholipid of the amyloplast membrane and are preserved from further breakdown by complexing with amylose molecules[59].

Similarly to the presence of starch lipids, small amount of proteins can be found in starch granules. The crude protein content typically accounts for <0.4% of the dry weight of starch[58], which includes granule internal and surface proteins. The internal proteins have been recognised as the enzymes participating in the starch synthesis process (GBSSI-II) with the surrounding granule proteinaceous matrix for the source of surface-associated proteins[28]. In terms of the protein molecular weight, M, the internal proteins are characterised as M~60-149 kDa, and M~5-30 kDa for the surface associated proteins respectively[60].

Minerals, along with the lipids and proteins, contribute to the minor compounds of starch granules. Phosphorus has attracted great attention due to its significant effects on the properties of starch[61]. The total phosphorus content of starch is generally ~0.03% and this ranges from 0.002-0.09% amongst species[62]. Additionally, the predominant form of phosphorus existing in starch varies among species. In cereal starches, phospholipids account for the majority of the phosphorus content[63-65], whereas amylopectin phosphate monoester is dominant for root/tuber starches; particularly for potato starch[62].

In terms of the relationship between the composition and digestive enzyme stability of starch, studies on the digestibility of starches from the same species show that starch granules with larger diameters are more resistant to hydrolysis than smaller granules. This is probably due to the lower surface area to volume ratio of larger starch granule[66-68]. The effect of granule size on the enzymatic stability of starch also reflects on the fact that starches from different botanical sources are hydrolysed to different extents by α-amylase: in the order wheat>maize>pea>potato[69]. This tendency is in line with the general concept that legume/root/tuber starches show more resistance to α-amylase hydrolysis than do cereal starches[70-72]. It also has been indicated that the enzymatic stability of native starch can be ascribed to the granule size only in part; composition and architecture play roles too[70-73].

Apart from the granule size effects associated with the enzymatic stability, starches with different amylose contents also exhibit different susceptibility to amylases. The extent of hydrolysis of cereal starches is inversely correlated with the amylose content[74-77]. This is associated with the existence of lipids with which amylose is complexed and (thus) decreases its accessibility to the α-amylase[72,78,79]. Additionally, the existence of amylose-lipid complexes tends to limit the granule from swelling/hydration and thus the potential penetration and degradation by α-amylases[73,79,80]. For tuber/root starches, an inverse correlation between the amylose content and digestibility has been observed among starches from different varieties of potato and yam[81,82]. As mentioned previously, unlike cereal starches, tuber/root starches are essentially free from lipids (<0.1). Hence, the apparent effect of amylose content may reflect the interplay between the granule molecular structure, crystallinity and hence enzymatic stability of the starch.

4 MOLECULAR STRUCTURE

Amylopectin has a very high molecular weight which ranges typically from 10^7-10^9 Daltons[61]. It has, as mentioned above, a highly branched structure where ~5% of the α-glucosidic bonds are α(1-6) linked as 'branch points'. Several hypotheses with respect to its structural conformation, such as 'laminated' structure[83], 'comb' structure[84], 'tree-like' structure[85], have been postulated over many years. Some of these are presented in Figure 1. Amongst these progressive models, the 'tree-like' conformation has been favoured. This is because the multiple branch points within the model for internal unit chains are compatible with the enzymatic evidence for the structural properties of amylopectin[86]. The 'tree-like' conformation, as a prototype, has been further developed into the cluster conformation[87]. The unit chains of amylopectin vary between DP 6-100, occasionally up to DP 200 or more. The majority of the chains, however, are between DP 6-35, and hence its average chain length is defined as DP 23-44[88].

The amylopectin unit chains can be categorised into three different 'classes' depending on their association. The A-chains are the chains linked to other unit chains via their reducing ends by α(1-6) linkages but not branched by other unit chains. The B-chains carry A- and B-chains, and their reducing ends are connected to other B-chains or the single C-chain. The C-chain also carries other chains but is the only reducing-end containing chain per amylopectin molecule. In terms of their spatial orientation, the A-chains are accommodated within a single cluster, while the B-chains are interspersed between clusters. Depending upon the number of clusters in which B-chains are located, they can be subcategorised into B_1- to B_4-chains[89].

The average chain lengths of A-chains for amylopectins from different species are in the range DP 12-16; DP 20-24 for B_1-chains; DP 42-48 for B_2-chains; DP 69-75 for B_3-chains and DP 100+ for B_4-chains[89]. Amongst different botanical origins, cereal amylopectins are generally the shortest in terms of chain lengths (especially short chains), followed by root and tuber starches[90,91]. A ratio of A- to B-chains also varies amongst different species. A ratio of 2.2 has been reported for waxy rice, 1.5 for tapioca, 1.2 for maize and 0.8 for potato[89,92]. However, these ratios vary depending on the literature source and the means of analysis[93]. In addition to the A-/B-chains ratio, the ratio between short chains (A- and B_1-chains; external chains) and long chains (B_{2+}-chains; internal chains) is also species-dependent. Waxy rice plus waxy and normal maize starches are characterised by a ratio of ~10-11:1; 8:1 for tapioca starch and 6:1 for potato starch[88]. By estimating the difference between the lengths of B_2- and B_3-chains, a general length of an entire cluster has been suggested to be approximately 27-28 glucosyl units, which is equivalent to 9.5-9.8 nm (~0.35 nm per glucosyl unit), regardless of the species[89].

The exact spatial orientation of amylopectin branch points, in parallel with that of the unit chains, within the cluster model of amylopectin has also received attention. The location of branch points has been suggested to be accommodated in the less crystalline

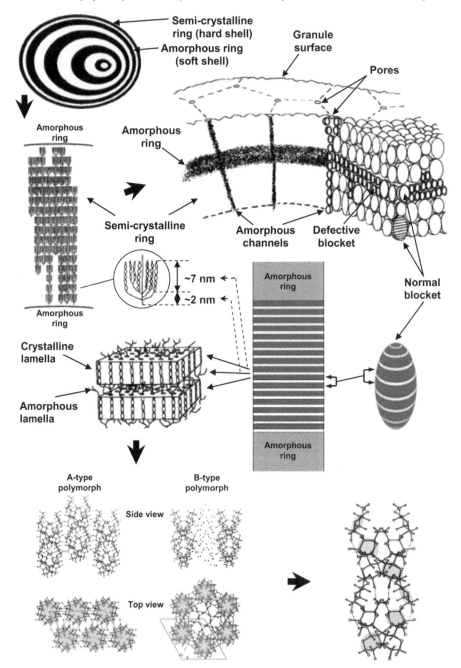

Figure 1 *Overview of starch granule structure redrawn from Gallant et al. (1997)[134] showing the schematic drawings of the arrangements of amylopectin clusters and both the amorphous and crystalline lamellae within a semi-crystalline growth ring[116]. The original blocklet structure is superimposed with the features of the defective blocklets forming amorphous channels and the heterogeneous shells with the normal blocklets[13]*

(amorphous) regions of starch since any remnants after lintnerisation (with acid) consist predominantly of either linear (DP 15) or single branched (DP 25) molecules[94]. Further investigations on the location of branch points of amylopectin suggest that the distribution of them between the crystalline regions and the amorphous regions differs according to species[95]. The acid-hydrolysis remnants of maize (including its waxy counterpart) and tapioca starches consist of most singly branched unit chains, followed in branching extent by banana, then potato and high amylose maize starches. This observation indicates that there are more branch points that are embedded in or close to the crystalline regions of maize and tapioca starches compared to banana, potato and high amylose maize starches. Moreover, as shown by enzymatic approaches (α-amylolysis), maize (*wx* and *wxdu*) starches have a higher density of branch points in the amorphous regions than do potato and *aewx* maize starches[96-98].

Amylose, compared to amylopectin, is of a lower molecular weight. Its average molecular weight ranges from $1.1 \times 10^5 - 6.5 \times 10^5$ Da on a number average basis (M_n), $3.2 \times 10^5 - 1.1 \times 10^6$ Da on a weight average basis (M_w)[91]. The molecular weight of amylose, analogous to amylopectin, also varies among species. Generally, amylose molecules of cereal starches are characterised by a relatively low molecular weight compared to the root and tuber species. The M_n and M_w of cereal amylose molecules are in the ranges 1.1×10^5 - 2.0×10^5 and 2.9×10^5 - 7.9×10^5, respectively: $M_n = 2.5 \times 10^5$ - 6.8×10^5 and $M_w = 6.2 \times 10^5$ - 1.1×10^6 for root/tuber amylose molecules.

The structure of amylose was once considered to be a linear polymer because of its complete conversion to maltose by β-amylase. With improvements for purifying β-amylase free from α-amylase, it has been demonstrated that amylose is only partially hydrolysable (~70% for potato amylose) by β-amylase[99]. The incomplete degradation of amylose indicates the existence of α(1-6) linkages in its chemical structure. Further evidence suggests that amylose consists of linear and branched molecules[100]. The molar percentage of branched molecules ranges from ~30% - 70% among different species; cereal amylose generally containing less branch molecules (~30% - 44%) than tuber and root amylose (~34% - 70%)[101]. In terms of the number-average DP, the linear fractions of cereal amylose molecules are in the range 390-1270; 1000-2440 for branched molecules. In contrast, root and tuber amylose molecules are characterised by high molecular weights for both linear (870-3940) and branched (1980-6940) fractions[91,101-107]. Additionally the branched fractions of root and tuber amylose molecules consist of more branch linkages than that of cereal amylose (excluding wheat amylose). The branched fraction with more branch points, together with the high molar percentage, results in the amylose molecules (taken as a whole) of root and tuber starches having high number of chains (3.8 -12.2) compared to that of cereal starches (2.5 – 6.5).

As mentioned previously, when starch is subjected to α-amylase hydrolysis, the hydrolysis extent is decreased with an increase in the amylose content of starch. This is particularly true for cereals due to the effect of amylose-lipid complexes. These complexes are true inclusion complexes of cereal fatty acids and lysophospholipids within amylose single helices. With respect to the molecular structure, studies have revealed that the susceptibility to hydrolysis is correlated positively with the quantity of A-chains (DP 8-12) but negatively with B_1-chains (DP 18-23)[108]. Of interest, analysis of starch remnants has also shown that there is no preferential hydrolysis of the fraction of either amylose or amylopectin[71,109]. This can be confirmed by time-dependent hydrolysis (Figure 2) where amylose (debranched fraction 1; DF1) and amylopectin are hydrolysed simultaneously (in parallel with the hydrolysis extent of the starch). The consistent unselective α-glucan

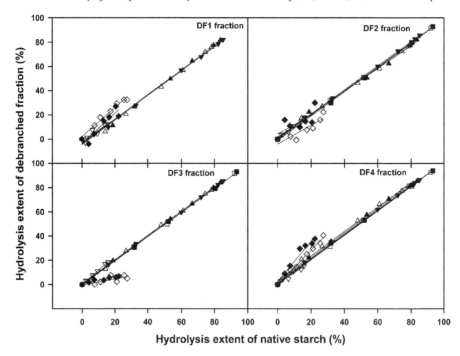

Figure 2 *First order regression analysis of the extent of hydrolysis (by α-amylase) of debranched fractions (DF1-4) compared to the extent of hydrolysis of the corresponding native starches: Amioca (■), waxy maize (□), maize-N (▲), maize-C (△), tapioca (▼), potato (▽), Hylon VII (◆) and amylomaize (◇)*

Table 1 *Comparison of the slopes (from Figure 2) from the first order regression analyses of the extent of hydrolysis (by α-amylase) of debranched fractions (DF1-4) compared to the extent of hydrolysis of the corresponding native starches*

Starch	DF1	DF2[b]	DF3	DF4
Amioca	—[a]	0.987	0.990	1.017
Waxy maize	—	0.995	0.991	1.012
Maize-N	0.957	0.971	1.008	1.045
Maize-C	0.940	0.975	1.005	1.067
Tapioca	0.947	1.009	1.003	1.026
Potato	0.765	1.045	1.180	0.967
Hylon VII	1.032	1.022 *	0.322	1.810
Amylomaize	1.294	0.579 *	0.235 *	1.413

[a] below the detection limit

[b] the r value for first order regression > 0.95, unless otherwise marked (*, 0.90 < r < 0.95).

hydrolysis is also evident for the unit chains of amylopectin (DF2, B_{2+} chains; DF3, B_1 chains; DF4, A chains) within the same starch with time.

The amylose and amylopectin molecules of starches, however, show different extents of α-amylase susceptibility: amylopectin > amylose for normal maize, tapioca and potato starches, but the opposite trend is true for high amylose maize starches (Table 1). In terms of the unit chains, $A > B_1 > B_{2+}$ is evident for waxy and normal maize starches; $A > B_{2+} > B_1$ for tapioca and high amylose maize and; $B_1 > B_{2+} > A$ for potato. The difference in the extent of α-amylase susceptibility can be related to the orientation of amylose in the crystalline lamellae and the unit chains of amylopectin by which the crystalline 'blocks' are constructed (Figure 1).

5 CRYSTALLINE LAMELLAE and POLYMORPHISM

Native starch exists as semi-crystalline granules in which the crystalline regions are formed by the double helices arranged in a more or less ordered fashion. The double helices are created mostly by the exterior chains of amylopectin molecules. They are thermodynamically favoured to associate with adjacent chains; probably as the result of the existence of branch points in the cluster structure[110,111]. The single chains of amylopectin double helices are paired in a parallel arrangement with a spatial shift of c/2 along the chain axis[112]. This spatial shift enables the double helixes to form ordered arrays where this arrangement is reinforced by the inter-double helical hydrogen bonds of O_2-O_6 and O_3-O_3. Approximately 57% of the double helices participate in the formation of crystallites where the double helical content of starches is 39-48% and 20-27% for crystallinity (as observed by solid-state nuclear magnetic resonance and X-Ray powder diffraction techniques)[113]. Due to the cluster structure of amylopectin, crystalline lamellae alternate with the amorphous lamellae arising from the branching points of amylopectin. This alternating arrangement can be characterised further by small-angle X-Ray (or neutron) diffractometry[114].

The repeat distance for crystalline and amorphous lamellae has been reported to be ~9 nm, which coincides with the size of a cluster based on chromatography (Figure 1)[115,116]. The ~9 nm repeat distance per set of crystalline and amorphous lamellae (ranging from 8.6 to 9.6 nm) and ~20 repeats (in average) per single stack of lamellar 'sets' appear to be nearly consistent for different starches[116-118]. The overall thickness of ~20 repeats is similar to that a growth ring (~120-400 nm) estimated by scanning/transmission electron microscopy[117]. However, the thickness of the actual crystalline lamellae are species-dependent and in the range 5.9-7.5 nm with a descending order of high-amylose pea > amylomaize > normal pea > barely/tapioca/low-amylose pea > wheat/maize/potato > waxy barley/rice > waxy maize[116-118]. For comparison, the thickness of amorphous lamellae range from 1.3-3.3 nm; the high-amylose pea starch amorphous lamellae are the smallest then followed by barley/amylomaize, potato/low-amylose pea/normal pea, tapioca/rice/maize/waxy barley/wheat, and waxy maize[116-118].

When starch is subjected to wide-angle X-Ray diffractometry, the particular arrangement of crystallite-constituting double helices results in the starch generating species-dependent diffraction patterns. The traditional concept is that cereal starches produce "A-type" diffraction patterns; whereas we find "B-type" for tuber and high amylose starches; and "C-type" for legume, root and some fruit and stem starches[111]. The A-type polymorph is characterised by a unit cell consisting of six double helices packed in a monoclinic fashion ($a = 2.124$ nm, $b = 1.172$ nm and $c = 1.069$ nm) with four water molecules situated in between the helices[112]. The integrity of the unit cell is maintained by the hydrogen bonds which are formed directly between the surrounding double helices or via the embedded water molecules. In contrast, the B-type form is less compact where double helices are arranged in a hexagonal fashion with dimensional parameters of $a = b = 1.85$ nm and $c = 1.04$ nm[119]. Additionally the central open core contains thirty six water molecules and forms a water channel. The well-hydrated unit cell makes the B-type polymorph able to be converted to the A-type after for example heat-moisture treatment, but the reverse conversion is probably not possible without disrupting the crystalline structure[110]. With respect to the C-type polymorph, it is now believed that the polymorphic conformation is the combination of A- and B-type polymorphs within the granule rather than an individual (granule) structure[120]. Additionally it has been suggested that the B-type structure is rich in the centre of granule and decreases progressively outwards[121,122].

In terms of the relationship between the polymorphism and digestive enzyme stability, studies on the susceptibility of *in vitro* synthesised spherulites from linear α-glucans (DP20) suggest that the spheroidal crystals with B-type polymorph are more resistant to α-amylase hydrolysis than the A-type[123,124]. This has also been confirmed using Lintner starches which show a lack of amorphous regions and most of the branch points when compared to native starches[124]. In the case of native starches, it has been reported that starches from tubers, legumes and high amylose crops are characterised by higher α-amylase resistance than cereals and those starches with higher resistance are characterised by B-type polymorphic structure. The predominant regulator in the susceptibility of starch to digestive enzymes, therefore, is often been ascribed to its polymorphism. Nevertheless it has also been pointed out that other factors, such as granule size, composition (including the extent of damage), molecular and crystalline structures need to be considered in conjunction[72,73,124].

6 α-AMYLASE EROSION PROFILE OF STARCH GRANULES

The particular structures of starch granules regulate their digestive patterns by α-amylases[125]. The surface of granules where enzymatic erosion commences is different amongst starches. From observations of the surface morphology of granules after α-amylase hydrolysis, it has been reported that any "holes" are distributed randomly for waxy and normal maize starches but are localised along the equatorial groove for wheat starch granules. The holes on the periphery of potato starch granules are also not geographically focused and much less in terms of number compared to that of waxy/normal maize starches[126]. In contrast, the location of hydrolysis is more concentrated on the edge of granules for sweet potato starch[127]; on one hemisphere for yam starch[127]; and far from the hilum for lotus rhizome starch[128]. These amylase-generated holes have been associated with surface pores and internal channels of native starch granules which facilitate penetration of α-amylase into granules[125,129,130].

The erosion profile of starch granules by α-amylases, in addition to the difference in the location where hydrolysis is initiated, is affected by the starch origin. The pattern of centripetal erosion has often been reported for waxy/normal maize and wheat starches of which numerous holes are formed on the surface. Progressively these are enlarged, forming external grooves penetrating towards the centre[126,131,132]. In contrast, a pattern of centrifugal corrosion is observed for potato and high amylose maize starches where a few holes are produced on the periphery. These lead to the granule core, whereupon the granules are hydrolysed to a greater extent by the α-amylase but in a diffusion manner[72,132,133]. An overall process for both erosion profiles of starch granules by pancreatic α-amylases has been postulated where they (i) diffuse onto the granules randomly, (ii) start the hydrolysis at these contact points, (iii) penetrate from the surface towards the centre of granules, then finally (iv) cause diffusion-regulated degradation which spreads gradually[72].

7 CONCLUSION

Starch stability in what are largely aqueous environments is achieved by dense packing with structural order. In the case of native granules, the α-glucans are associated into helices and crystalline regions as energy dense structures. However, these can be degraded by amylases to generate the energy for growth and nutritional needs. Without these stable structures, energy deposition of α-glucans would be far more difficult.

References

1 M. G. James, K. Denyer and A. M. Myers, *Curr. Opin. Plant Biol.*, 2003, **6**, 215.
2 M. Thitisaksakul, R. C. Jiménez, M. C. Arias and D. M. Beckles, *J. Cereal Sci.*, 2012, **56**, 67.
3 P. D. Commuri and P. L. Keeling, *Plant J.*, 2001, **25**, 475.
4 D. Delvallé, S. Dumez, F. Wattebled, I. Roldán, V. Planchot, P. Berbezy, P. Colonna, D. Vyas, M. Chatterjee, S. Ball, Á. Mérida and C. D'Hulst, *Plant J.*, 2005, **43**, 398.
5 F. D. Macdonald and J. Preiss, *Plant Physiol.*, 1985, **78**, 849.
6 J. L. Ozbun, J. S. Hawker and J. Preiss, *Plant Physiol.*, 1971, **48**, 765.
7 J. Craig, J. R. Lloyd, K. Tomlinson, L. Barber, A. Edwards, T. L. Wang, C. Martin, C. L. Hedley and A. M. Smith, *Plant Cell*, 1998, **10**, 413.
8 M. K. Morell, B. Kosar-Hashemi, M. Cmiel, M. S. Samuel, P. Chandler, S. Rahman, A. Buleon, I. L. Batey and Z. Li, *Plant J.*, 2003, **34**, 173.
9 G. J. Abel, F. Springer, L. Willmitzer and J. Kossmann, *Plant J.*, 1996, **10**, 981.
10 J. Marshall, C. Sidebottom, M. Debet, C. Martin, A. M. Smith and A. Edwards, *Plant Cell*, 1996, **8**, 1121.
11 I. Roldán, F. Wattebled, M. Mercedes Lucas, D. Delvallé, V. Planchot, S. Jiménez, R. Pérez, S. Ball, C. D'Hulst and Á. Mérida, *Plant J.*, 2007, **49**, 492.
12 N. Fedoroff, S. Wessler and M. Shure, *Cell*, 1983, **35**, 235.
13 O. E. Nelson, P. S. Chourey and M. T. Chang, *Plant Physiol.*, 1978, **62**, 383.
14 C.-Y. Tsai, *Biochem. Genet.*, 1974, **11**, 83.
15 B. Delrue, T. Fontaine, F. Routier, A. Decq, J. M. Wieruszeski, N. Van Den Koornhuyse, M. L. Maddelein, B. Fournet and S. Ball, *J. Bacteriol.*, 1992, **174**, 3612.
16 M. L. Maddelein, N. Libessart, F. Bellanger, B. Delrue, C. D'Hulst, N. Van den Koornhuyse, T. Fontaine, J. M. Wieruszeski, A. Decq and S. Ball, *J. Biol. Chem.*, 1994, **269**, 25150.
17 F. Wattebled, A. Buléon, B. Bouchet, J.-P. Ral, L. Liénard, D. Delvallé, K. Binderup, D. Dauvillée, S. Ball and C. D'Hulst, *Eur. J. Biochem.*, 2002, **269**, 3810.
18 K. Denyer, B. Clarke, C. Hylton, H. Tatge and A. M. Smith, *Plant J.*, 1996, **10**, 1135.
19 K. Denyer, D. Waite, S. Motawia, B. L. Møller and A. M. Smith, *Biochem. J.*, 1999, **340**, 183.
20 A. M. Kram, G. T. Oostergetel and E. F. J. van Bruggen, *Plant Physiol.*, 1993, **101**, 237.

21 Y. Takeda, H.-P. Guan and J. Preiss, *Carbohydr. Res.*, 1993, **240**, 253.

22 S. Ball, H.-P. Guan, M. James, A. Myers, P. Keeling, G. Mouille, A. Buléon, P. Colonna and J. Preiss, *Cell*, 1996, **86**, 349.

23 M. G. James, D. S. Robertson and A. M. Myers, *Plant Cell*, 1995, **7**, 417.

24 G. Mouille, M.-L. Maddelein, N. Libessart, P. Talaga, A. Decq, B. Delrue and S. Ball, *Plant Cell*, 1996, **8**, 1353.

25 Y. Nakamura, T. Umemoto, Y. Takahata, K. Komae, E. Amano and H. Satoh, *Physiol. Plant.*, 1996, **97**, 491.

26 D. Pan and O. E. Nelson, *Plant Physiol.*, 1984, **74**, 324.

27 A. Rahman, K.-s. Wong, J.-l. Jane, A. M. Myers and M. G. James, *Plant Physiol.*, 1998, **117**, 425.

28 A. M. Smith, *Curr. Opin. Plant Biol.*, 1999, **2**, 223.

29 S. C. Zeeman, T. Umemoto, W.-L. Lue, P. Au-Yeung, C. Martin, A. M. Smith and J. Chen, *Plant Cell*, 1998, **10**, 1699.

30 J. Kossmann, G. J. W. Abel, F. Springer, J. R. Lloyd and L. Willmitzer, *Planta*, 1999, **208**, 503.

31 C.-T. Larsson, P. Hofvander, J. Khoshnoodi, B. Ek, L. Rask and H. Larsson, *Plant Sci.* 1996, **117**, 9.

32 A. Edwards, D. C. Fulton, C. M. Hylton, S. A. Jobling, M. Gidley, U. Rössner, C. Martin and A. M. Smith, *Plant J.*, 1999, **17**, 251.

33 J. R. Lloyd, V. Landschütze and J. Kossmann, *Biochem. J.*, 1999, **338**, 515.

34 W. R. Morrison, T. P. Milligan and M. N. Azudin, *J. Cereal Sci.*, 1984, **2**, 257.

35 R. F. Tester, S. J. J. Debon and J. Karkalas, *J. Cereal Sci.*, 1998, **28**, 259.

36 A. L. Charles, Y.-H. Chang, W.-C. Ko, K. Sriroth and T.-C. Huang, *Starch/Stärke*, 2004, **56**, 413.

37 I. Defloor, I. Dehing and J. A. Delcour, *Starch/Stärke*, 1998, **50**, 58.

38 K. Alvani, X. Qi, R. F. Tester and C. E. Snape, *Food Chem.*, 2011, **125**, 958.

39 M. Yusuph, R. F. Tester, R. Ansell and C. E. Snape, *Food Chem.*, 2003, **82**, 283.

40 T. Nakamura, M. Yamamori, H. Hirano, S. Hidaka and T. Nagamine, *Mol. Gen. Genet.*, 1995, **248**, 253.

41 H. Ceballos, T. Sánchez, N. Morante, M. Fregene, D. Dufour, A. M. Smith, K. Denyer, J. C. Pérez, F. Calle and C. Mestres, *J. Agric. Food Chem.*, 2007, **55**, 7469.

42 A. E. McPherson and J. Jane, *Carbohydr. Polym.*, 1999, **40**, 57.

43 M. Yamamori, S. Fujita, K. Hayakawa, J. Matsuki and T. Yasui, *Theor. Appl. Genet.*, 2000, **101**, 21.

44 H. Ceballos, T. Sánchez, K. Denyer, A. P. Tofiño, E. A. Rosero, D. Dufour, A. Smith, N. Morante, J. C. Pérez and B. Fahy, *J. Agric. Food Chem.*, 2008, **56**, 7215.

45 S. Gomand, L. Lamberts, R. Visser and J. Delcour, *Food Hydrocolloids*, 2010, **24**, 424.

46 G. P. Schwall, R. Safford, R. J. Westcott, R. Jeffcoat, A. Tayal, Y.-C. Shi, M. J. Gidley and S. A. Jobling, *Nat. Biotechnol.*, 2000, **18**, 551.

47 T.-j. Lu, J.-l. Jane, P. L. Keeling and G. W. Singletary, *Carbohydr. Res.*, 1996, **282**, 157.

48 R. F. Tester and J. Karkalas, *Starch/Stärke*, 2001, **53**, 513.

49 Y.-C. Shi, P. A. Seib and J. E. Bernardin, *Cereal Chem.*, 1994, **71**, 369.

50 R. Tester, W. Morrison, R. Ellis, J. Piggot, G. Batts, T. Wheeler, J. Morison, P. Hadley and D. Ledward, *J. Cereal Sci.*, 1995, **22**, 63.

51 J. E. Cottrell, C. M. Duffus, L. Paterson and G. R. Mackay, *Phytochemistry*, 1995, **40**, 1057.

52 S. J. J. Debon, R. F. Tester, S. Millam and H. V. Davies, *J. Sci. Food Agric.*, 1998, **76**, 599.

53 R. F. Tester, S. J. Debon, H. V. Davies and M. Gidley, *J. Sci. Food Agric.*, 1999, **79**, 2045.
54 M. Asaoka, J. M. V. Blanshard and J. E. Rickard, *Starch/Stärke*, 1991, **43**, 455.
55 D. G. Medcalf and K. A. Gilles, *Cereal Chem.*, 1965, **42**, 558.
56 N. U. Haase and J. Plate, *Starch/Stärke*, 1996, **48**, 167.
57 J. C. Shannon, D. L. Garwood and C. D. Boyer, in *Starch: Chemistry and technology*, ed. J. N. BeMiller and R. L. Whistler, Academic Press, San Diego, 3rd edn., 2009, ch. 3, pp. 23–82.
58 J. J. M. Swinkels, *Starch/Stärke*, 1985, **37**, 1.
59 W. R. Morrison, *J. Cereal Sci.*, 1988, **8**, 1.
60 P. M. Baldwin, *Starch/Stärke*, 2001, **53**, 475.
61 R. F. Tester and J. Karkalas, in *Polysaccharides II: Polysaccharides from Eukaryotes*, ed. E. J. Vandamme, S. De Baets and A. Steinbüchel, Wiley-VCH, 2002, vol. 6, ch. 13, pp. 381–438.
62 T. Kasemsuwan and J.-L. Jane, *Cereal Chem.*, 1996, **73**, 702.
63 W. R. Morrison, R. F. Tester, C. E. Snape, R. Law and M. J. Gidley., *Cereal Chem.*, 1993, **70**, 385.
64 R. F. Tester and W. R. Morrison, *Cereal Chem.*, 1990, **67**, 551.
65 W. R. Morrison, R. V. Law and C. E. Snape, *J. Cereal Sci.*, 1993, **18**, 107.
66 T. Noda, S. Takigawa, C. Matsuura-Endo, T. Suzuki, N. Hashimoto, N. Kottearachchi, H. Yamauchi and I. Zaidul, *Food Chem.*, 2008, **110**, 465.
67 C. A. Knutson, U. Khoo, J. E. Cluskey and G. E. Inglett, *Cereal Chem.*, 1982, **59**, 512.
68 T. Noda, T. Kimura, M. Otani, O. Ideta, T. Shimada, A. Saito and I. Suda, *Carbohydr. Polym.*, 2002, **49**, 253.
69 S. G. Ring, J. M. Gee, M. Whittam, P. Orford and I. T. Johnson, *Food Chem.*, 1988, **28**, 97.
70 J.-l. Jane, Z. Ao, M. Duvick, Susan A.and Wilklund, S.-H. Yoo, K.-S. Wong and C. Gardner, *J. Appl. Glycosci.*, 2003, **50**, 167.
71 H. W. Leach and T. J. Schoch, *Cereal Chem.*, 1961, **38**, 34.
72 R. F. Tester, X. Qi and J. Karkalas, *Anim. Feed Sci. Technol.*, 2006, **130**, 39.
73 R. F. Tester, J. Karkalas and X. Qi, *World Poultry Sci. J.*, 2004, **60**, 186.
74 H. Fuwa, M. Nakajima, A. Hamada and D. V. Glover, *Cereal Chem.*, 1977, **54**, 230.
75 T. Vasanthan and R. S. Bhatty, *Cereal Chem.*, 1996, **73**, 199.
76 H. Fuwa, N. Inouchi, D. V. Glover, S. Fujita, M. Sugihara, S. Yoshioka, K. Yamada and Y. Sugimoto, *Starch/Stärke*, 1999, **51**, 147.
77 J. Li, T. Vasanthan, R. Hoover and B. Rossnagel, *Food Chem.*, 2004, **84**, 621.
78 W. R. Morrison, R. F. Tester, M. J. Gidley and J. Karkalas, *Carbohydr. Res.*, 1993, **245**, 289.
79 M. Lauro, P. M. Forsell, M. T. Suortti, S. H. D. Hulleman and K. S. Poutanen, *Cereal Chem.*, 1999, **76**, 925.
80 J. Karkalas, R. F. Tester and W. R. Morrison, *J. Cereal Sci.*, 1992, **16**, 237.
81 C. K. Riley, A. O. Wheatley, I. Hassan, M. H. Ahmad, E. Y. S. A. Morrison and H. N. Asemota, *Starch/Stärke*, 2004, **56**, 69.
82 M. E. Karlsson, A. M. Leeman, I. M. Björck and A.-C. Eliasson, *Food Chem.*, 2007, **100**, 136.
83 W. N. Haworth, E. L. Hirst and F. A. Isherwood, *J. Chem. Soc.*, 1937, 577.
84 H. Staudinger and E. Husemann, *Justus Liebigs Ann. Chem.*, 1937, **527**, 195.
85 K. H. Meyer, *Cell. Mol. Life Sci.*, 1952, **8**, 405.
86 D. J. Manners and N. K. Matheson, *Carbohydr. Res.*, 1981, **90**, 99.
87 D. French, *J. Jpn. Soc. Starch Sci.*, 1972, **19**, 8.

88 S. Hizukuri, *Carbohydr. Res.*, 1985, **141**, 295.

89 S. Hizukuri, *Carbohydr. Res.*, 1986, **147**, 342.

90 J. Jane, Y. Y. Chen, L. F. Lee, A. E. McPherson, K. S. Wong, M. Radosavljevic and T. Kasemsuwan, *Cereal Chem.*, 1999, **76**, 629.

91 S. Hizukuri, J.-i. Abe and I. Hanashiro, in *Carbohydrates in Food*, ed. A.-C. Eliasson, CRC Press, 2006, ch. 9, pp. 305–390.

92 S.-H. Yun and N. K. Matheson, *Carbohydr. Res.*, 1993, **243**, 307.

93 D. J. Manners, *Carbohydr. Polym.*, 1989, **11**, 87.

94 J. P. Robin, C. Mercier, R. Charbonniere and A. Guilbot, *Cereal Chem.*, 1974, **51**, 389.

95 J.-l. Jane, K.-S. Wong and A. E. McPherson, *Carbohydr. Res.*, 1997, **300**, 219.

96 E. Bertoft, *Carbohydr. Res.*, 1991, **212**, 229.

97 C. Gérard, V. Planchot, P. Colonna and E. Bertoft, *Carbohydr. Res.*, 2000, **326**, 130.

98 Q. Zhu and E. Bertoft, *Carbohydr. Res.*, 1996, **288**, 155.

99 S. Peat, S. J. Pirt and W. J. Whelan, *J. Chem. Soc.*, 1952, 705.

100 W. Banks, C. T. Greenwood and J. Thomson, *Makromol. Chem.*, 1959, **31**, 197.

101 Y. Takeda, S. Hizukuri, C. Takeda and A. Suzuki, *Carbohydr. Res.*, 1987, **165**, 139.

102 S. Hizukuri, Y. Takeda, M. Yasuda and A. Suzuki, *Carbohydr. Res.*, 1981, **94**, 205.

103 S. Hizukuri, *J. Jpn. Soc. Starch Sci.*, 1988, **35**, 185.

104 S. Hizukuri, Y. Takeda, N. Maruta and B. O. Juliano, *Carbohydr. Res.*, 1989, **189**, 227

105 C. Takeda, Y. Takeda and S. Hizukuri, *Cereal Chem.*, 1989, **66**, 22.

106 K. Shibanuma, Y. Takeda, S. Hizukuri and S. Shibata, *Carbohydr. Polym.*, 1994, **25**, 111.

107 Y. Yoshimoto, T. Takenouchi and Y. Takeda, *Carbohydr. Polym.*, 2002, **47**, 159.

108 S. Srichuwong, T. C. Sunarti, T. Mishima, N. Isono and M. Hisamatsu, *Carbohydr. Polym.*, 2005, **60**, 529.

109 P. Colonna, A. Buléon and F. Lemarié, *Biotechnol. Bioeng.*, 1988, **31**, 895.

110 A. Imberty, A. Buléon, V. Tran and S. Péerez, *Starch/Stärke*, 1991, **43**, 375.

111 R. F. Tester, J. Karkalas and X. Qi, *J. Cereal Sci.*, 2004, **39**, 151.

112 A. Imberty, H. Chanzy, S. Pérez, A. Bulèon and V. Tran, *J. Mol. Biol.*, 1988, **201**, 365.

113 D. Cooke and M. J. Gidley, *Carbohydr. Res.*, 1992, **227**, 103.

114 J. Blanshard, D. Bates, A. Muhr, D. Worcester and J. Higgins, *Carbohydr. Polym.*, 1984, **4**, 427.

115 R. Cameron and A. Donald, *Polymer*, 1992, **33**, 2628.

116 P. J. Jenkins, R. E. Cameron and A. M. Donald, *Starch/Stärke*, 1993, **45**, 417.

117 P. J. Jenkins and A. M. Donald, *Int. J. Biol. Macromol.*, 1995, **17**, 315.

118 P. A. Perry and A. M. Donald, *Int. J. Biol. Macromol.*, 2000, **28**, 31.

119 A. Imberty and S. Pérez, *Biopolymers*, 1988, **27**, 1205.

120 B. Pfannemüller, *Int. J. Biol. Macromol.*, 1987, **9**, 105.

121 T. Y. Bogracheva, V. J. Morris, S. G. Ring and C. L. Hedley, *Biopolymers*, 1998, **45**, 323.

122 A. Buléon, C. Gérard, C. Riekel, R. Vuong and H. Chanzy, *Macromolecules*, 1998, **31**, 6605.

123 G. Williamson, N. J. Belshaw, D. J. Self, T. R. Noel, S. G. Ring, P. Cairns, V. J. Morris, S. A. Clark and M. L. Parker, *Carbohydr. Polym.*, 1992, **18**, 179.

124 V. Planchot, P. Colonna and A. Buleon, *Carbohydr. Res.*, 1997, **298**, 319.

125 J. E. Fannon, J. A. Gray, N. Gunawan, K. C. Huber and J. N. BeMiller, *Cellulose*, 2004, **11**, 247.

126 D. Gallant, C. Mercier and A. Guilbot, *Cereal Chem.*, 1972, **49**, 354.

127 J.-C. Valetudie, P. Colonna, B. Bouchet and D. J. Gallant, *Starch/Stärke*, 1993, **45**, 270.

128 H.-M. Lin, Y.-H. Chang, J.-H. Lin, J.-l Jane, M.-J. Sheu and T.-J. Lu, *Carbohydr. Polym.*, 2006, **66**, 528.

129 J. E. Fannon, R. J. Hauber and J. N. BeMiller, *Cereal Chem.*, 1992, **69**, 284.

130 J. E. Fannon, J. M. Shull and J. N. BeMiller, *Cereal Chem.*, 1993, **70**, 611.

131 V. Planchot, P. Colonna, D. J. Gallant and B. Bouchet, *J. Cereal Sci.*, 1995, **21**, 163.

132 W. Helbert, M. Schülein and B. Henrissat, *Int. J. Biol. Macromol.*, 1996, **19**, 165.

133 E. Sarikaya, T. Higasa, M. Adachi and B. Mikami, *Process Biochem.*, 2000, **35**, 711.

134 D. J. Gallant, B. Bouchet and P. M. Baldwin, *Carbohydr. Polym.*, 1997, **32**, 177.

135 H. Tang, T. Mitsunaga and Y. Kawamura, *Carbohydr. Polym.*, 2006, **63**, 555.

3

ENYMATIC DEGRADATION OF CELL WALL POLYSACCHARIDES

G.A. Tucker

University of Nottingham, School of Biosciences, Sutton Bonington LE12 5RD, UK
gregory.tucker@nottingham.ac.uk

1 INTRODUCTION

The plant cell wall represents the major source of carbohydrate polymers on the planet. Indeed it has been estimated that cellulose and hemicelluloses (in particular xylans) represent the two most abundant polysaccharides in nature. It is not surprising therefore that cell wall derived polysaccharides are commonly exploited for commercial means. Nature has generated a wide range of structural polysaccharides that are associated with the cell wall. Some of these can be exploited in their natural form but their functionality can often be enhanced by structural modifications. These modifications can be carried out post extraction of the polymers from the plant. Alternatively, understanding how the plant synthesizes and metabolises these polymers may lead to *in planta* methods for modification. The major *in planta* modifications are often associated with the degradation (or turnover) of the wall polymers during specific stages of plant growth and development. This chapter will firstly consider the general structure of the plant cell wall and the enzymatic capacities that result in degradation *in planta*. This will be followed by a more detailed consideration of the changes that can be brought about in the structure of pectin during fruit ripening. This provides an illustration of how modifications to cell wall degrading enzymes can result in changes in polymer structure.

2 THE PLANT CELL WALL

Plant cells differ from those in many other eukaryotic organisms in that they possess a cell wall. The cell wall serves many purposes within the plant primarily it provides mechanical strength to the plant tissues, either through turgor pressure or the deposition of woody material, it thus determines the size and shape of individual cells and this in turn dictates the overall structure of the plant. The wall can also serve as a means of defence against attack by microbes. The nature of the wall can be highly heterogenous[1] as will be seen also in the following chapter in this volume by Grassby *et al*. In growing tissues this wall needs to be able to extend to accommodate cell expansion and as such is a dynamic structure. In these tissues the wall is termed a primary wall. The primary cell wall is typically composed of 90% carbohydrate and 10% structural protein. There are two types of primary cell wall found in plants. Type I is common in dicotolydenous plants and this includes most fruit and vegetables. The general structure of this wall is described by Carpita and Gribeault[2].

Type II cell walls are found in monocots such as the major cereal crops maize and wheat and the general structure has again been described by Carpita[3]. The two types of wall both contain cellulose which can account for around 30% of the total wall mass. Cellulose is composed of β(1-4) linked glucose residues which form linear polymers that can vary between 5000 and 25000 residues in length. In the plant cell wall these polymers are aggregated together to form crystalline structures called fibrils. Each fibril consists of approximately 36 cellulose polymers, laid down in a parallel fashion and held together by hydrogen bonding[4].

These fibrils interact with hemicellulosic polymers. There are a wide range of hemicelluloses polymers[5]. Hemicelluloses in general are thought to be able to hydrogen bond with the surface of the cellulose fibril and in some cases may even penetrate the fibril and disrupt its microcrystalline structure. Also in some cases the hemicellulose may span the gap between fibrils and serve to generate a cellulose/hemicellulose framework serving to anchor the fibril into the wall.

The major hemicellulose in type I cell walls is xyloglucan. This polymer consists of a backbone of β(1-4) linked glucose residues with the majority of these residues associated with a xylose "side group". Further substitution of the xylose residues, with galactose and fucose, occurs and this again can be variable. In type II walls the major hemicelluloses polymers are based on xylans and these form, after cellulose, the second most abundant polysaccharide on Earth. The majority of "xylans" in land plants comprise a β(1-4) linked xylan backbone with a number of side chain substitutions. Arabinoxylans represent a major hemicellulose in the starchy endosperm of several crops and can account for 0.15% of the rice grain and up to 13% of whole grain flour from barley and 30% of wheat bran. This polymer has a linear backbone of xylose residues part substituted with arabinose on either the O-2 or O-3 positions. Some of these arabinose residues are further esterified to ferulic or coumaric acids. Another major difference between the two cell wall types is in their pectin composition. Type I primary cell walls can contain up to 30% pectin and this polymer constitutes the matrix between the cellulose fibrils along with the structural protein components of the wall. In comparison type II cell walls tend to have much smaller amounts, around 10%, of pectin. The structure of pectin is covered in section 3 below.

In mature tissues, especially those found in the woody parts of the plant, these have ceased expansion and are now primarily involved in providing structural rigidity. The cell walls in these tissues are similar to the primary cell walls but have extra layers which are highly lignified and are referred to as secondary cell walls. Lignin is a generic term for an aromatic polymer that is formed from 4-hydroxyphenyl propanoid building blocks. The general structure of lignin can be seen in Freudenberg[6] and a more recent review is that by Vanholme *et al.*[7]. These building blocks are normally derived from the three monolignols-coniferyl, sinapyl and coumoryl alcohols. These units are coupled together through oxidative reactions into a heterologous polymer that serves to strengthen the wall and resist microbial attack. The secondary wall is thus a relatively inert and stable tissue. In contrast the primary cell wall is a dynamic structure and undergoes significant metabolism and turnover to accommodate growth and development of the cell and tissue. This involves a large number of enzymes that are capable of modifying or degrading the various cell wall polymers. The cellulose fibril, being in a highly organized crystalline state, is relatively inert with respect to these degradative enzymes, although putative cellulase genes have been identified in many plants, thus it is the hemicellulose and pectin fractions that appear to undergo the major modifications during plant development. This chapter will focus on modifications to pectin structure, in particular during tomato fruit ripening.

3 PECTIN STRUCTURE

Pectin is a major structural component of many plant cell walls. Pectin is primarily composed of galacturonic acid residues and consists of three major components, namely homogalacturonan, rhamnogalacturonan I and rhamnogalacturonan II[8].

Homogalacturonan consists of long chains of α(1-4) linked galacturonic acid residues. These can exist as either the free acid or may be methyl esterified at the C6 position. The degree of methyl esterification (DE) is a major functional parameter of this polymer. The extent and distribution of de-esterified groups within the polymer may have a distinct impact on the functionality of the polymer within the wall. Thus a run of de-esterified residues could be envisaged to strengthen the wall through the chelation of calcium ions to form the so called "egg box" junctions between polymers. Alternatively, de-esterified regions of pectin could act as sites for polygalacturonase action resulting in pectin depolymerisation and wall loosening. The extent and distribution of these de-esterified regions may be controlled through the action of pectinesterase (PE) enzymes. Additionally the galacturonic acid residues may also be acetylated or in certain polymers may be linked to xylose to form xylogalacturonan.

Rhamnogalacturonan 1 consists of an alternating rhamose/galacturonic acid backbone which may be several hundred residues long. The rhamnose residues can be further decorated by side chains of galactose and arabinose residues. The structure and distribution of these side chains can vary considerably between polymers. In contrast the third pectin polymer-rhamnogalactronan-II- appears to be highly conserved between plants but represents only a small proportion of the total pectin in the wall.

4   CELL WALL DEGRADING ENZYMES

Plants have the potential to synthesise a very wide range of cell wall modifying enzymes. The *Arabidopsis* genome would appear to contain around 730 open reading frames representing potential glycoside hydrolases and glycosyl transferases[9], the two major classes of enzymes involved in cell wall metabolism. The glycosyl transferases are thought to be primarily involved in cell wall synthesis and the hydrolases in degradation or turnover, and we focus here on the hydrolase and related enzymes thought to be implicated in cell wall degradation. A review by Fry[10] provides a good overview of the range of plant enzymes involved in this cell wall metabolism. Fischer and Bennett[11] provide a good review of those specifically involved in fruit ripening. Most studies on hemicellulose metabolism have concentrated on the modification of xyloglucan. The three major enzyme activities involved are probably endo-glucanase (EGase), xyloglucan endotransglycosylase (XET) and expansin. The endo-glucanase potentially acts by breaking β(1-4) linked bonds. Whilst these could reside within a cellulose polymer the more likely substrate is the backbone of the xyloglucan[12], although precise modes of action of these enzymes are unclear. The XET can function to modify the xyloglucan by breaking a β(1-4) linked bond and then reforming this by addition to either an adjacent polymer or to a short oligosaccharide[13]. This enzyme is sometimes bi-functional and can act simply as a xyloglucan hydrolase. Expansin is unique among the proteins described here in that it is not a hydrolase, indeed does not actually catalyse a chemical reaction at all. The mode of action of expansin is that the protein can intercede between the surface of the cellulose fibril and the hemicellulose breaking the hydrogen bonds that hold these polymers together and thus allowing slippage of the fibril in relation to the hemicelluloses[14]. The silencing of expansin expression in tomato fruit has been shown to slow down softening[15].

There are a wide range of enzymes that can modify pectin in the cell wall the major ones being pectinesterase (PE), polygalacturonase (PG) and galactanase. Pectinesterase removes the methyl group from the C6 of a galacturonic acid to generate a carboxylic acid moiety. This action can be random, as with the enzyme from many fungal sources, or in the plant itself may result in a block of de-esterified galacturonic acids. Polygalacturonase cleaves the bond between two adjacent de-esterified galacturonic acids thus resulting in the depolymerisation of the pectin backbone and potentially increasing the solubility of the pectin. Many plants contain a related enzyme activity- pectin or pectate lyase which can also act to depolymerise the pectin but does so through a β-elimination reaction that does not always require the presence of de-esterified galacturonic acids at the site of cleavage. Plants also contain exo-galactanse enzyme activities one potential role for which is to modify the galactan side chains found within the rhamnogalacturonan 1.

In addition there are many other hydrolytic enzymes that have been found associated with plant cell walls, these include xylanases, arabinosidases, mannosidases and many others. For a comprehensive list of carbohydrate metabolising enzymes the reader is referred to the carbohydrate-active enzymes database (CAZY) WEB site at http://www.cazy.org

5 PECTIN METABOLISM IN RIPENING TOMATO FRUIT

Softening is a major factor determining the shelf life of tomato fruit. Softening may be brought about by a combination of factors, but the most important is likely to be the degradation of the cell wall that accompanies ripening. The changes in pectin structure during ripening of tomato have been well documented and are common to those found in many other fruit[11, 16-18]. Thus during ripening the pectin undergoes increased solubilisation, depolymerisation of the soluble fraction, a reduction in the DE and a loss of neutral sugars in particular galactose. Pectin is supposedly secreted into the cell wall with a DE of around 100%. During tomato fruit development this declines to a value of about 80% by the breaker stage with a further decline to about 40% during ripening.

A number of enzymes have been found in tomato fruit that are likely to be able to degrade the cell wall *in vivo*, those targeted at pectin include polygalacturonase (PG), a hydrolytic enzyme that results in pectin depolymerisation, pectinesterase (PE) that can reduce the degree of esterification and β-galactosidase which can remove galactosyl residues. All have been targets for genetic manipulation in particular by the application of gene silencing technology[18,19]. This approach has demonstrated the role of these specific enzymes in the depolymerisation, deesterification and loss of galactose. Two groups have independently down regulated PG activity in tomato fruit[20, 21]. This resulted in a reduction in PG activity to below 1% of that normally detected and the depolymerisation of the pectin occurring during normal ripening was markedly inhibited[22]. Transgenic fruit handled better with less cracking[23] and this enabled fruit to be harvested from the vine later in ripening and as such with an improved flavour. A commercial version of this fruit was marketed in the USA under the trade name "Flavr Savr". Another commercial advantage was that puree made from the transformed fruit had improved quality[23-25] and this was again commercialised. Despite these commercial advantages the transgenic fruit were only slightly firmer than controls[24, 26, 27]. The lack of effect on texture could have been explained by the fact that whilst PG activity had been reduced to around 1% of normal this still represents a significant level of enzyme activity given the unusually high level of this enzyme found in tomato fruit. More recently, Cooley and Yoder[28] produced a line in which the PG gene was disrupted by a transposon insertion. This resulted in the almost complete elimination of PG activity but did not appear to result in any further reduction in firmness.

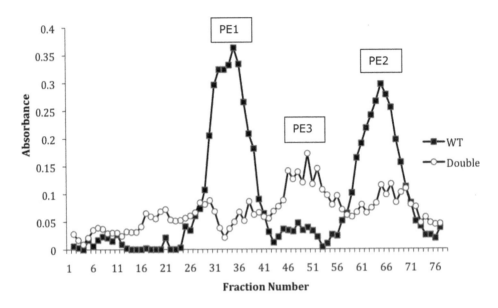

Figure 1 *Pectinesterase isoform profile. Total PE activity was extracted from pericarp tissue of mature green* Ailsa craig *(WT) or genetically modified tomato fruit (Double) containing antisense genes to PE1 and PE2. PE isoforms were separated by heparin affinity chromatography*

β-galactosidase (β-gal) has been shown to occur in tomato fruit in at least three isoforms but only one of these, β-gal II, is thought to be capable of degrading cell wall pectin[29, 30]. Thus, β-gal II has been the target for gene silencing. At least seven putative β-galactosidase genes (*TBG 1-7*) have been identified in tomato fruit[31, 32] and, of these, *TBG4* may be the most significant gene for the expression of cell wall directed galactanase activity associated with tomato fruit ripening. Antisense suppression of this gene has resulted in a 90% reduction in exo-galactanase activity in transgenic fruit[32], and this can be associated with a 40% increase in the firmness of the ripe fruit[33].

In tomato fruit PE activity can be separated, by heparin affinity chromatography, into at least three isoforms, namely PE1, PE2 and PE3[34]. A typical isoform profile of activity from mature green fruit is shown in Figure 1. The two major isoforms PE1 and PE2 are present at similar levels at this stage of ripening but PE2 appears to become more dominant in the ripe fruit. Genes for both these isoforms have been identified and then subjected to gene silencing. The effect on total PE activity in ripe ("Breaker +8") fruit, in each case, is shown in Figure 2. The effect on the degree of esterification in each case is shown in Figure 3. Silencing of isoform 1 resulted in only a moderate change in either the total enzyme activity in ripe fruit or the observed DE pattern but did result in an enhanced softening of the fruit[35]. This might indicate that PE1 is involved in strengthening the wall during fruit development.

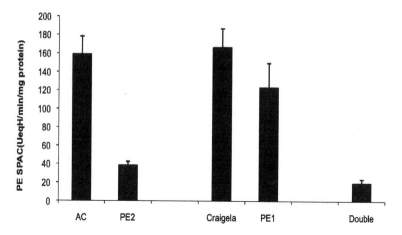

Figure 2 *Pectinesterase activity in wild type and antisense tomato fruit. Total PE was extracted from B+8 tomato pericarp from wild type* Ailsa craig *(AC) or Craigella fruit or equivalent lines carrying antisense genes for PE2 or PE1, respectively. Activity from a line in which both PE1 and PE2 have been silenced is shown as "Double"*

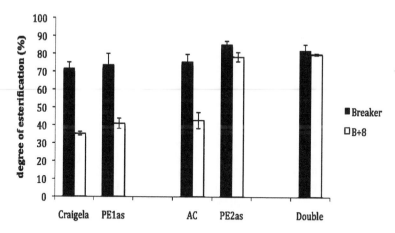

Figure 3 *Pectin de-esterification patterns in wild type and antisense tomato fruit. Total pectin was extracted from Breaker and B+8 tomato pericarp from wild type* Ailsa craig *(AC) or Craigella fruit or equivalent lines carrying antisense genes for PE1 (PE1as) or PE2 (PE2as), respectively. Pectin de-esterification from a line in which both PE1 and PE2 have been silenced is shown as "Double"*

Two groups of researchers have independently down regulated the expression of PE2 activity[36, 37]. In both cases expression of PE activity was reduced to around 20% of that in normal fruit. The degree of esterification, in ripe fruit, was between 15 and 40% higher in transgenic fruit compared to controls[36,37] and this was accompanied by a reduced depolymerisation of the polyuronides in ripe fruit and a decrease in the amount of chelator soluble pectin of around 20-30% [36]. These changes in pectin structure had little or no effect on fruit firmness during normal ripening[38]. However, they were associated with a complete loss of tissue integrity in over-ripe fruit[38] and with an increase in quality of processed juice made from the transgenic fruit, these having higher soluble solids and viscosity[39]. This suggests that sites for polygalacturonase action may pre-exist in green fruit and that there is limited scope for synergy until well into the ripening process.

More recently crosses between these two antisense lines have generated plants in which both of these isoforms have been simultaneously silenced[40]. The isoform profile from these double antisense fruit is shown in Figure 1 for comparison with the wild type profile, and the impact on total PE activity and pectin de-esterification in Figures 2 and 3, respectively. It can be seen that fruit from these plants, in which both genes are silenced, still express 5-10% residual PE activity (Figure 2) apparently associated with one or more minor isoforms (Figure 1). It would appear from Figure 1 that the activity associated with PE2 is entirely lost in these double antisense fruit whilst that associated with PE3 would appear to be unaffected. The peak of activity associated with PE1 is much reduced but there is evidence of some residual activity suggesting that either the antisense has not been totally effective or that this peak in wild type fruit is the product of more than one isoform. Studies examining the salt dependency of PE activity associated with this PE1 isoform peak would tend to support the latter hypothesis and that this peak does indeed represent the combined activity of at least two independent isoforms of PE[40]. The DE of the pectin in these fruit still undergoes a reduction to 80% at the breaker stage indicating either a role for these minor isoforms in this decrease or the fact that the pectin is not 100% esterified on secretion. The impact on fruit softening is not major[40] but there was a tendency for the fruit to soften faster that controls, similar to the effect of anti-sensing PE1 individually[35].

References

1   R.A. Burton, M.J. Gidley and G.B. Fincher, *Nature Chem. Biol.*, 2010, **6**, 724.
2   N.C. Carpita and D.M. Gibeault, *The Plant Journal*, 1993, **3**, 1.
3   N.C. Carpita (1996) *Ann. Rev. Plant Physiol. & Plant Mol. Biol.* 1996, **47**, 445.
4   Y. Nishiyama, *J. Wood Science*, 2009, **55**, 241.
5   H.V. Scheller and P. Ulvskov, *Ann. Rev. Plant Biol.*, 2010, **61**, 263.
6   K. Freudenberg, *Science,* 1965, **148**, 595.
7   R. Vanholme, B. Demedts, K. Morreel, J. Ralph and W. Boerjan, *Plant Physiol.*, 2010, **153**, 895.
8   W.G.T. Willats, L. McCartney, W. Mackie and P. Knox, *Plant Mol. Biol.,* 2001, **47**, 9.
9   B. Henrissat, P.M. Coutinho and G.J. Davies, *Plant Molecular Biology*, 2001, **47**, 55.
10 S.C. Fry, *Ann. Rev. Plant Physiol. and Plant Mol. Biol.*, 1995, **46**, 497.
11 R.L. Fischer and A.B. Bennett, *Ann. Rev. Plant Physiol. and Plant Mol. Biol.*, 1991, **42**, 675.
12 D.A. Brummell, M.H. Harpster, P.M. Civello, J.M. Palys, A.B. Bennett and P. Dunsmuir, *The Plant Cell* 1999, **11**, 2203.
13 J. de Silva, D. Arrowsmith, A. Heelyer, S. Whiteman and S. Robinson, *J. Exp. Botany*, 1994, **45**, 1693.
14 S.J. McQueen-Mason and D.J. Cosgrove, *Plant Physiol.,* 1995, **107**, 87.

15 D.A. Brummell, B.D. Hall and A.B. Bennett, *Plant Mol. Biol.* 1999, **40,** 615.

16 G.A. Tucker, in *The Biochemistry of Fruit Ripening*, ed. G.B. Seymour, J.E. Taylor and G.A. Tucker, Chapman and Hall, 1993, p.1.

17 J. Giovannoni *Ann. Rev. Plant Physiol. and Plant Mol. Biol.*, 2001, **52,** 725.

18 D.A. Brummell and M.H. Harpster, *Plant Mol. Biol.* 2001, **47,** 311.

19 G. Tucker, in *Texture in Foods, Volume 2, Solid Foods*, ed. D. Kilcast, Woodhead Publishing Ltd., 2004, p. 321.

20 R.E. Sheehey, M. Kramer and W.R. Hiatt, *Proc. Nat. Acad. Sci. USA*, 1988, **85,** 8805.

21 C.J.S. Smith, C.F. Watson, J. Ray, C.J. Bird, P.C. Morris, W. Schuch and D. Grierson, *Nature*, 1988, **334,** 724.

22 C.J.S. Smith, C.F. Watson, P.C. Morris, C.R. Bird, G.B. Seymour, J.E. Gray, C. Arnold, G.A. Tucker, W. Schuch, S.E. Harding and D. Grierson, *Plant Mol. Biol.*, 1990, **14,** 369.

23 W. Schuch, J. Kanczler, D. Robertson, G.E. Hobson, G.A. Tucker, D. Grierson, S. Bright and C. Bird, *Horticult. Sci.*, 1991, **26,** 1517.

24 M. Kramer, R. Sanders, H. Bolkan, C. Waters, R.E. Sheehy and W.R. Hiatt, *Postharvest Biol. Tech.*, 1992, **1,** 241.

25 D.A. Brummell and J.M. Labovitch. *Plant Physiol.*, 1997, **115,** 717.

26 K.R. Langley, A. Martin, R. Stenning, A.J. Murray, G.E. Hobson, W. Schuch and C.R. Bird, *J. Sci. Food Agric.,* 1994, **66,** 547.

27 N. Errington, G.A. Tucker and J.R. Mitchell., *J. Sci. Food Agric.*, 1998, **76,** 515.

28 M.B. Cooley and J.I. Yoder, *Plant Mol. Biol.*, 1998, **38,** 521.

29 R. Pressey, *Plant Physiol.,* 1983, **71,** 132.

30 A. Carey, K. Holt, S. Picard, R. Wilde, G.A. Tucker, C.R. Bird, W. Schuch and G.B. Seymour, *Plant Physiol.,* 1995, **108,** 1099.

31 D.L. Smith, D.A. Starrett and K.C. Gross KC, *Plant Physiol.,* 1998, **117,** 417.

32 D.L. Smith and K.C. Gross, *Plant Physiol.,* 2000, **123,** 1173.

33 D.L. Smith, J.A. Abbott and K.C. Gross, *Plant Physiol.,* 2002, **129,** 1755.

34 G.A. Tucker, N.G. Robertson and D. Grierson, *J. Sci. Food Agric.,* 1982, **33,** 396.

35 T.D. Phan, W. Bo, G. West, G.W. Lycett and G.A. Tucker, *Plant Physiol.,* 2007, **144,** 1960.

36 D.M. Tieman, R.W. Harriman, G. Ramamohan and A.K. Handa, *The Plant Cell*, 1992, **4,** 667.

37 L.H. Hall, G.A. Tucker, C.J. Smith, C.F. Watson G.B. Seymour, Y. Bundick, J.M. Boniwell, J.D. Fletcher, J.A. Ray, W. Schuch, C.R. Bird and D. Grierson., *The Plant Journal*, 1993, **3,** 121.

38 D.M. Tieman and A.K. Handa, *Plant Physiol.,* 1994, **106,** 429.

39 B.R. Thakur, R.K. Singh, D.M. Tieman and A.K. Handa, *J. Food Sci.,* 1996, **61,** 85.

40 W. Bo and G.A. Tucker (unpublished observations)

4

FUNCTIONAL COMPONENTS AND MECHANISMS OF ACTION OF 'DIETARY FIBRE' IN THE UPPER GASTROINTESTINAL TRACT: IMPLICATIONS FOR HEALTH

T. Grassby*, C.H. Edwards, M. Grundy, P.R. Ellis

Biopolymers Group, Diabetes and Nutritional Sciences Division, School of Medicine, King's College London, Waterloo Campus, London, SE1 9NH, UK
*terri.grassby@kcl.ac.uk

1 INTRODUCTION

Dietary fibre (DF) is a generic term for a chemically diverse group of carbohydrates that are resistant to endogenous enzymes of the human digestive tract, and includes various non-starch polysaccharides (both natural and synthetic), starches and short-chain oligosaccharides. This short review will focus on the structure and properties of naturally-occurring non-starch polysaccharides (NSP), specifically in the form of native cell wall polymer networks and also cell wall fractions of variable composition. The pioneering epidemiological studies of researchers, such as Trowell and Burkitt, helped to stimulate early research on DF, showing that fibre had much broader potential health benefits, notably for heart disease, diabetes, and cancer, than previously thought.[1] Recent reviews have examined the physiological effects and health benefits associated with consumption of DF in more detail.[2-5] For instance, NSP are now known to reduce the rate and extent of digestion and absorption of macronutrients, such as starch and lipid,[6-8] which leads to attenuation of postprandial glycaemia, lipaemia and also fasting blood cholesterol.[9-11] Such metabolic effects are linked to a reduced risk of developing coronary heart disease, stroke, obesity and type 2 diabetes mellitus. The mechanisms by which NSP are thought to alter the digestion kinetics and other gut functions are diverse, including formation of viscous solutions, encapsulation of nutrients, inhibition of digestive enzymes, and binding of bile salts. Furthermore, fermentation of NSP by microorganisms in the large intestine results in the production of short chain fatty acids (SCFA), which are now considered to have important positive effects on gut function and metabolism.[12,13] This review will focus only on the mechanisms of NSP that are considered to alter the time course of digestion and postprandial metabolism.

1.1 Definition of Dietary Fibre

The term crude fibre was in common usage prior to the 1950s. It was only defined by the analytical method used to measure it, namely that portion of plant food lost during ashing, after hydrolysis by acid and alkali,[14] but the majority of the cell wall is destroyed by this chemical treatment, giving artificially low results that are not nutritionally relevant.[15] The first publication of the term 'dietary fibre' is generally attributed to Hipsley,[16] although he did not define exactly what he meant by this term. Trowell first defined it as, *"the skeletal remains of plant cells that are resistant to digestion by enzymes of man"*.[17] He later expanded his definition to include, *"plant polysaccharides and lignin resistant to hydrolysis by the digestive enzymes of man"*,[18] which meant that lignin and some storage polysaccharides were also included.

Ever since, the definition and analysis of DF has been a subject of great debate and controversy, and two detailed reviews on the history of this are recommended.[19,20] Only recently has partial agreement been reached on the definition adopted by the Codex Alimentarius Commission (Table 1):

Table 1 *Definition of dietary fibre as adopted by the Codex Alimentarius Commission (ALINORM 09/32/26 and 10/33/26)[21, 22]*

"Dietary fibre means carbohydrate polymers[a] with ten or more monomeric units[b], which are not hydrolysed by the endogenous enzymes in the small intestine of humans and belong to the following categories:
  i. Edible carbohydrate polymers naturally occurring in the food as consumed,
  ii. carbohydrate polymers, which have been obtained from food raw material by physical, enzymatic or chemical means and which have been shown to have a physiological effect of benefit to health as demonstrated by generally accepted scientific evidence to competent authorities,
  iii. synthetic carbohydrate polymers which have been shown to have a physiological effect of benefit to health as demonstrated by generally accepted scientific evidence to competent authorities

[a]When derived from a plant origin, dietary fibre may include fractions of lignin and/or other compounds associated with polysaccharides in the plant cell walls. These compounds also may be measured by certain analytical method(s) for dietary fibre. However, such compounds are not included in the definition of dietary fibre if extracted and re-introduced into a food.
[b]Decision on whether to include carbohydrates of 3 to 9 monomeric units should be left up to national authorities."

However, the current CODEX definition still has issues that need to be resolved; notably whether or not it should include fibre from animal sources, lignin, resistant starch, oligosaccharides with 3-9 monomeric units, and the requirement for proof of physiological benefit for added fibre.[23]

1.2 Types of Dietary Fibre

The CODEX definition of DF, while useful for the food industry, is not specific enough for scientific research, which should specify the type of fibre being investigated, preferably having been well characterised. In nutrition research, DF is often described as being 'water-soluble' or 'water-insoluble'. While it may be possible to define the solubility of a fibre *in vitro*, it is very difficult to determine the rate and extent of dissolution in the gut environment, where factors such as pH, mixing conditions, polymer concentration, and ionic conditions vary significantly. The following terms are preferable as they are more specific and are of more use when searching the literature (Table 2):

Table 2 *Types of dietary fibre with selected examples*

Non-starch polysaccharides (NSP)	Resistant starch (RS)	Short-chain carbohydrates (DP>3)
Cell wall NSP	RS1 – Physically	Galacto-
Cellulose	inaccessible	oligosaccharides
Hemicelluloses:	(encapsulated)	(GOS)
Xylan	starch	
Xyloglucan (XG)		Fructo-
Arabinoxylan (AG)		oligosaccharides
Mannan		(FOS)
Glucomannan	RS2 – Native	
Galactomannan (GM)	starch	Inulin
β-glucan		
Callose		Raffinose
Pectins:		
Homogalacturonan (HG)	RS3 – Retrograded starch	Polydextrose
Rhamnogalacturonan I		Stachyose
Rhamnogalacturonan II		
Arabinan		Maltodextrins
Galactan	RS4 – Modified	
Arabinogalactan I	starches	
Arabinogalactan II		
Storage NSP		
Guar (galactomannan)		
Other NSP		
Psyllium seed husk mucilage (arabinoxylan)		

Non-starch polysaccharides are the main components of native plant cell walls. The cell wall NSP are divided into the following categories: cellulose, hemicelluloses and pectic materials. Hemicelluloses and pectins can be further subdivided depending on their monomeric composition and glycosidic linkages (see section 2). Storage polysaccharides, such as guar galactomannan (GM), which is found as the main polysaccharide in the endospermic cell walls of the Indian Cluster bean (*Cyamposis tetragonoloba* (L.) Taub), are also included in this category. Some of these NSP, including guar GM, are cold water-soluble and form highly viscous solutions even at low polymer concentrations ($\leq$ 1% w/w).[24]

Resistant starch (RS) is the fraction of starch that is not digested in the upper intestine and is therefore included in the current definition of DF. It can be divided into the following categories: RS1 – starch that is physically inaccessible to enzymes; RS2 – starch that is not fully gelatinised, due to insufficient heat or water present during processing and/or cooking; RS3 – starch that has been allowed to cool after gelatinisation, forming retrograded starch; or RS4 – starch that has been chemically modified to resist digestion. Resistant starches may be fermented by the microbiota in the large intestine to produce SCFA.[25] RS4 will not be discussed in this review.

Short chain carbohydrates, including resistant oligosaccharides (ROS), are low molecular weight carbohydrates, which contain 3-10 monosaccharide units, and are usually located as intracellular constituents of edible plant tissues. They are generally separated from associated polysaccharides on the basis of their solubility in 80% (v/v) ethanol,[26] although this method does not necessarily exclude all ROS with more than 10 monosaccharide units.[27] The monosaccharides in ROS are linked by β-glycosidic bonds that are resistant to human gastrointestinal enzymes; however ROS can be degraded by colonic bacteria to produce SCFA. ROS are either extracted from edible plants, produced by partial enzyme hydrolysis of xylan, inulin, or starch, or synthetically by the reverse reaction from disaccharides.[28,29]

1.3 Common Methods of Analysis for Dietary Fibre Content

Accurate and reproducible measurements of the DF content of foods are needed not only for legislative and food labelling purposes, but also for researchers designing test diets for human and animal studies. This information is, not surprisingly, valuable to the food industry in supporting health claims, and to policy makers developing healthy eating campaigns. Ideally, a single method would quantify each DF component. A historical perspective of the development of fibre analysis methods has been reported elsewhere.[19] Recently, a method has been developed, validated and adopted by AOAC International (the Association of Analytical Communities) that quantifies DF as high or low molecular weight, including RS and ROS. This covers the CODEX definition in one, albeit complicated, analysis (AOAC method 2009.01).[30] A further AOAC method (2011.25) separates the high molecular weight DF into water-soluble and insoluble fractions,[31] although it is not clear if this method is physiologically relevant for the reasons discussed above.

1.4 Physical and Chemical Conditions during Digestion

The digestive system (Figure 1) consists of the digestive tract (mouth, oesophagus, stomach, small intestine and colon), and the associated organs (salivary glands, liver, pancreas and gallbladder). Ingested food is masticated and mixed with saliva in the mouth, re-formed into a bolus, and then swallowed. The bolus travels down the oesophagus to the stomach, where it is exposed to gastric juices to form chyme. On passing out of the stomach, through the pyloric sphincter, the chyme is mixed with secretions from the pancreas, liver and gallbladder in the duodenum, and then moves along the small intestine

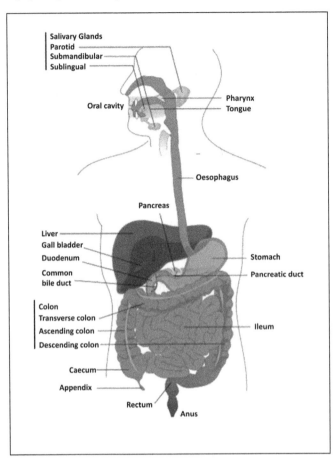

Figure 1 *The organs of the human digestive system*

into the colon. The colon is colonised by various symbiotic microorganisms that may further degrade and utilise undigested food components; these bacteria and any food residue are excreted as faeces. The process of digestion facilitates the breakdown of food and absorption of nutrients, such as starch, protein and lipid.[32] Normal digestive transit times are usually between 20 and 48 hours and are generally longer for women than for men.[33]

1.4.1 Physical Conditions. Mastication is a complex process of physical destruction of the edible plant tissue and the formation of a bolus that can be swallowed safely. The dimensions of the particles in the bolus will depend on the food and the individual, and may range from 5 µm to 3 cm. However, particle size tends to be reasonably consistent between individuals, for the same food, if measured at the swallowing threshold.[34-36] The ripeness and type of processing the plant food has undergone before ingestion may have significant effects on its behaviour during mastication. In some foods (e.g. legumes), the cooking process can weaken the bonds between adjacent cell walls, making cell separation during chewing more likely. Ripening may have a similar effect in some plant foods. In raw foods, and some cooked foods with unusual cell wall chemistry, such as Chinese water chestnuts, cell rupture is more common.[37] Whether the cells rupture or separate can have a significant impact on the immediate release of macronutrients in the proximal gut (Figure

2). Once the bolus is swallowed peristalsis pushes the food down the oesophagus into the stomach. Muscular contractions of the different gastric regions encourage mixing with gastric fluid; however, as shown by magnetic resonance imaging, the mixing is not sufficient to prevent sedimentation of solid particles.[38] Coordinated contractions of the lower region of the stomach, with a force of 0.65 N,[39] may grind the food to some degree before it passes through the pyloric sphincter into the duodenum. The effectiveness of this grinding is influenced by the compression strength of the food and the viscosity of the accompanying liquid.[39] In a heterogeneous meal, passage through the pylorus may initially favour liquids and small particles, a process known as 'gastric sieving'. The rate of gastric emptying is regulated to deliver ~8-16 kJ of energy to the duodenum per minute. This mechanism is dependent on many factors, including the nutrient and energy content of the food as well as the viscosity and volume of the chyme.[32]

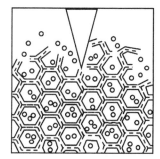 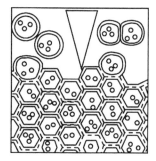

Figure 2 *Schematic of cell rupture (left) and cell separation (right). Redrawn from Brett and Waldron[37]*

In the small intestine, where typical transit times are in the range 3-5 hours, segmenting contractions mix the chyme, while peristaltic contractions transport it to the large intestine, the main part of which is the colon.[33] The transit time for the large intestine is considerably longer than that for the small intestine, allowing the gut microorganisms time to ferment some of the DF and other undigested components remaining in the food residue.

1.4.2 (Bio)chemical Conditions. Saliva mostly consists of water (99%), but also contains electrolytes (sodium, potassium, calcium, magnesium, bicarbonate, and phosphates), proteins (mucins, amylase and peroxidase) and organic molecules. Saliva is roughly neutral, pH 5.3-7.8, with buffering capacity supplied by bicarbonate, phosphates and urea.[40] The water and mucins in saliva are responsible for hydrating, lubricating and, in the case of plastic or deformable foods such as bread, binding together the bolus, thus making it easier to swallow. Starch is digested by salivary α-amylase into maltose, maltotriose and limit dextrins (activity 10-160 IU/ml).[41] In infants, saliva also contains lingual lipase,[42] however, it has not been detected in adults.[43]

Gastric juice has an acidic pH of 1-2.5, although due to poor mixing in the stomach and the inherent buffering capacity of food, the centre of a cohesive bolus may remain at its starting pH for some time.[39] Thus, salivary α-amylase may still be active inside the bolus. The acidic environment provides some protection against potentially harmful bacteria. A layer of mucus protects the stomach lining from the acids and enzymes that are present. Gastric juice also contains gastric lipase, and pepsin, which begin the digestion of triacylglycerols (TAG) and proteins, respectively. Gastric lipase digests 5-40% of the TAG present in the stomach, producing diacylglycerol (DAG) and free fatty acids, and a further 7.5% in the duodenum.[44, 45]

Once the chyme reaches the duodenum, it is neutralised to ~pH 6.5 by bicarbonate secreted from the pancreas and mucosa. As the chyme passes along the small intestine, the pH continues to gradually increase, reaching ~pH 7.5 at the terminal ileum.[46] Bile salts, pancreatic lipase, co-lipase, pancreatic amylase, trypsin and chymotrypsin are also secreted into the small intestine, facilitating further breakdown of lipid, starch and protein. Enzymic hydrolysis of starch to glucose is completed by the brush-border enzyme maltase, which is located in the intestinal mucosa, for absorption into the hepatic portal blood. The pancreatic lipase/co-lipase complex digests the lipids further, accounting for 40-70% of the overall lipid digestion. The products of lipid digestion are then incorporated into micelles that are absorbed across the intestinal mucosa.[45]

A variable fraction of the starch and other macronutrients remain undigested by the time the food reaches the terminal ileum. However, at this stage of digestion, as far as it is known, the DF components of plant food remain more or less physically and chemically intact. The proportion of nutrients that are released from the food matrix and potentially available for absorption is referred to as the bioaccessible fraction, whereas the term bioavailable only describes the nutrients that are actually absorbed.[47]

The undigested material reaches the colon, where DF is degraded by colonic microorganisms into SCFA, notably acetic, propionic and butyric acids. The majority of these SCFA are then absorbed by colonic epithelial cells and either metabolised *in situ* or transported, via the portal blood, to the liver where they are metabolised.[48]

2 PLANT CELL WALLS AND NON-STARCH POLYSACCHARIDES

Plant cell walls (PCW) consist of two or three distinct layers. The middle lamella forms during cell division and maintains contact between adjacent cells. It is relatively thin and contains mostly pectin. The primary wall, located between the middle lamella and the plasma membrane, is thicker and structurally more robust. It consists of complex interacting networks of cellulose, hemicelluloses and pectic material, supplemented with small amounts of structural proteins and, depending on species, phenolic-polysaccharide cross-links (Figure 3). The main structural components of the primary wall in most angiosperms (flowering plants) are usually cellulose, xyloglucan, pectins and extensin (Type I wall); whereas for the commelinid monocots, including grasses, they may include cellulose, arabinoxylan, β-glucan, xyloglucan and phenolic acids (Type II wall).[49]

In addition to the middle lamella and primary cell wall, some cells (xylem, phloem, epidermis tissues) develop a secondary cell wall. The interpolymeric spaces in secondary walls are filled by the indigestible phenolpropanoid polymer lignin, which makes them more hydrophobic and prevents further growth of the cell. Lignin confers additional structural stability to the walls, particularly when under compression. However, cells with secondary cell walls tend to make up a small proportion of the plant foods regularly consumed in a 'Western diet', as they are often removed to make the food more palatable (e.g. potato skins and wheat bran).[51]

PCW composition is very variable, with differences reported between species, tissues, cells, and even at the cellular level.[52,53]

2.1 Structural Plant Cell Wall Polysaccharides

In PCW chemistry the polysaccharides are commonly grouped into the following categories:

2.1.1 Cellulose. In most plant tissues, cellulose microfibrils form the scaffold around which the other PCW polysaccharides are assembled. Multiple chains of β(1→4)Glc are synthesised by rosettes of cellulose synthase modules embedded in the plasma membrane, and the chains then hydrogen bond to each other to form microfibrils. The microfibrils generally lie parallel to the surface of the plasma membrane and to each other within any given layer, but each layer has a different orientation.[54] This allows the PCW to withstand the turgor pressure from within the cell. The cellulose is prevented from forming one solid mass by the other cell wall components.[55]

2.1.2 Pectins. Pectins are usually defined empirically as those polysaccharides that are extracted from the wall by hot aqueous solutions of chelating agent (i.e. CDTA) or dilute acid. They have a wide variety of structures and compositions, but generally contain rhamnose, galactose, arabinose, galacturonic acid and glucuronic acid. Cereals usually have very little pectin, whereas in other plants, especially fruits, pectins can be the major polysaccharide constituent of the wall.

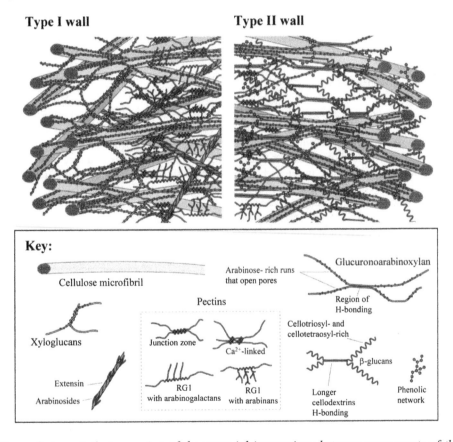

Figure 3 *An artist's impression of the potential interactions between components of the primary cell wall for Type I (most angiosperms) and Type II (commelinid monocots) cell walls. Not drawn to scale. Reproduced from Buchanan et al.[50] with permission of the American Society of Plant Biologists*

Some pectins, such as homogalacturonan (HG), form ordered gel-like structures via ionic bonds. Concerted ionic bonding between Ca^{2+} ions and negatively charged galacturonate residues of adjacent HG chains can create inter- and intra-chain bonds.[37] The acidic pH of the stomach may cause displacement of the Ca^{2+} ions by protonation of the carboxyl groups, potentially causing the HG chains to become disordered and therefore more water soluble. On the other hand, once the chains are fully protonated they are more likely to hydrogen bond to each other, which may reduce their solubility.[56]

As some pectins can be extracted from the PCW by hot aqueous solutions, plant materials lose some of their pectin content to the cooking water. The cell-cell adhesion is also weakened, due to β-elimination and methyl ester de-esterification of HG in the middle lamella, leading to cell separation during mastication.[57]

2.1.3 Hemicelluloses. Hemicelluloses are the PCW polysaccharides that are generally extracted using alkali solutions. For example, in cereals the major hemicelluloses are usually mixed-linkage β-glucans or arabinoxylans (AX), whereas in most other plants (e.g. legume seeds) they are usually xyloglucans (XG). Xylans and xyloglucans can hydrogen bond to the surface of cellulose microfibrils, and they were thought to provide the main connections between the cellulose microfibrils, but recent evidence suggests that the supramolecular structure is more complex than this.[58]

2.2 Other Structural Components

2.2.1 Ferulic acid and polysaccharide cross-linking by diferulic acids. Ferulic acid is a phenolic acid that is often found esterified to cell wall polysaccharides. Dimers of ferulic acid (diferulic acids) are formed *in vivo* via a peroxidase-mediated radical reaction, which can also lead to higher oligomers of ferulic acid.[59,60] Arabinoxylans in maize and bamboo, and rhamnogalacturonan I in sugar beet have been shown to have diferulic acid cross-links.[61-63] It is likely that these cross-links between the polysaccharide chains make them more resistant to solubilisation,[64] therefore maintaining the integrity of the cell walls during digestion. However, the presence of these linkages does not seem to impede utilisation of the PCW by the gut microbiota.[65] The ester bonds tethering the diferulic acids to the polysaccharides are susceptible to alkali hydrolysis, and this is how they are usually isolated from PCW.[66]

2.2.2 Cell wall proteins and enzymes. The cell wall proteins tend to be proteins that form part of the cell wall structure, or enzymes that modify the structure. Of the many structural proteins found in the wall, there is one that may have particular influence on wall structure as it relates to digestibility. Extensins can form intermolecular di-isodityrosine linkages (with the help of extensin peroxidase) to make a complex insoluble network which interpenetrates the other networks in the wall and interacts with pectin, possibly reducing the solubility of the pectin in the process.[67] Expansins are responsible for acid-mediated wall stress relaxation at pH 4.5-6, allowing cell growth without significant reduction in cell wall strength.[68]

The cell wall enzymes are able to modify the wall to allow cell growth and then further modify it during ripening, harvesting, storage and processing. During fruit ripening, multiple cell wall enzymes actively modify and/or degrade the cell wall polysaccharides.[69] This would change the cell wall properties and potentially the degree of cell-cell adhesion, thereby altering the composition of the DF. Post-harvest maturation of vegetables, such as asparagus spears, results in the deposition of cellulose and lignin in the cell walls, making solubilisation and digestion more difficult.[70] These changes could have implications for the bioaccessibility of the nutrients within the cell. The cell wall proteome is complex and not fully elucidated; the reader is directed to a review by Jamet *et al* for more information.[71]

2.3 Storage Polysaccharides

Storage polysaccharides, such as xyloglucans, galactans and mannans, are generally found in the cell walls of seeds, although some fructans and mannans are found in vacuoles.[72] They are laid down as the tissue matures, and then are mobilized again during germination as a source of nutrition.[73] A number of these storage polysaccharides are used in processed foods as bulking agents, thickeners, emulsifiers and stabilisers. Xyloglucans are present as storage polysaccharides in the cell walls of some tree seeds, particularly those of tamarind (*Tamarindus indica*). Konjac tubers (*Amorphophallus* sp.) yield a soluble glucomannan with 5-10% acetyl substitution of the mannan backbone. Storage galactomannans are usually identified as gums on food labels, and are obtained from a wide range of sources, including guar, fenugreek, locust bean and tara. Section 3.2 of this review will focus on guar gum, probably the most commonly studied storage polysaccharide.[74]

3 MECHANISMS OF DIETARY FIBRE IN RELATION TO ITS PHYSIOLOGICAL AND METABOLIC EFFECTS

In the past, when DF was investigated in prospective cohort studies, the participants' diets were generally classified into limited categories of foods and the importance of specific types of fibre was lost. The latest studies have recorded more detailed dietary data and investigated the effects of DF from specific plant food groups, namely cereals, legumes, vegetables, and fruits. These studies have shown that higher fibre intakes are associated with lower total mortality;[75,76] nevertheless, there are still serious deficiencies in our understanding of the biological mechanisms of DF. This is partly because of the experimental limitations of obtaining access to different sites of the human gut, and also because of the complexity and variations in structure and behaviour of cell walls and cell wall constituents in the gut environment. The following sections briefly review some of the work done on the mechanistic aspects of cell wall structures and NSP, in relation to physiological and metabolic effects.

3.1 Increased Viscosity

The viscosity of water-soluble NSP, such as guar GM, is known to strongly influence gastric emptying and sieving, intestinal transit time, and digesta flow and mixing.[24] The degree of viscosity generated in the gut by these soluble polysaccharides is dependent on the rate and extent of dissolution, and the structure, concentration and molecular weight of the polymer. Other factors, such as the presence of particulates, can also potentially influence the rheology of the digesta as observed in model systems of starch and NSP.[77,78] However, the rheology of gut digesta still remains poorly understood because (a) access to the digestive tract is difficult in humans (due to ethical and practical constraints), (b) there are methodological problems with reliably measuring digesta viscosity, and (c) it has complex flow behaviour.[79,80]

In order for NSP to increase the viscosity of the digesta, it must first hydrate and form a molecular dispersion. Determining whether NSP are water soluble or insoluble *in vivo* is not always predictable, although there are some features of NSP that can make them more or less likely to be soluble. Cellulose consists of linear chains of $\beta(1\rightarrow4)$ linked glucose, and is insoluble because significant portions of adjacent polymer chains hydrogen bond to each other forming a crystalline structure. Conversely, oat β-glucan, a polysaccharide of $\beta(1\rightarrow4)$ and $\beta(1\rightarrow3)$ linked glucose is generally soluble in water, because the random inclusion of the β(1→3) linkages gives the polymer a more disordered conformation. Also, branching of wheat AX and the charged groups of pectins prevent close packing, and therefore these polysaccharides tend to be more soluble.[81] Guar GM consists of a β(1→4) linked mannose backbone heavily substituted by galactose at the *O*-6 position, and therefore it is soluble in water because the galactose impedes contiguous interactions of the polymer chains. However, the hydration rate of guar GM is affected by many factors such as the concentration, molecular weight and particle size of the preparation.[82, 83] Therefore, in studies investigating the clinical efficacy and physiological mechanisms of other water-soluble NSP, the characterisation of the polysaccharide, such as molecular weight and dissolution behaviour, is crucial. However, these characteristics are not always reported by researchers, making comparisons between different studies extremely difficult.

The results of a series of randomised-controlled trials by Wood and colleagues, who carefully characterised the oat β-glucan in the food as eaten by the volunteers, indicated that cooking and processing can have significant effects on metabolic responses. Initially, they exposed muffins containing oat β-glucan to multiple freeze-thaw cycles and measured the properties of the β-glucan, and the muffins' effects on blood glucose responses relative to fresh muffins. Four freeze-thaw cycles reduced the solubility of β-glucan after *in vitro* digestion, slightly decreased the molecular weight and consequently adversely affected the peak blood glucose response and incremental area under the glucose response curve.[84] Following on from this study, the effects of molecular weight and solubility/extractability were investigated further by treating oat β-glucan with β-glucanase to give different molecular weight fractions (120 kDa – 2000 kDa) in the muffins. The solubility of the β-glucan from the muffins was variable with respect to molecular weight, but there was a significant inverse relationship between the product of the extractability and molecular weight of the β-glucan and the peak blood glucose rise found in 10 volunteers.[85]

The Wood group then investigated the effect of processing techniques on depolymerisation of β-glucan. Oat porridge and oat granola showed little depolymerisation, whereas oat pasta and oat crisp bread had molecular weight reductions of >80%. They concluded that depolymerisation occurred because of prolonged low temperature exposure to enzymes present in wheat flour, and that rapid denaturation of these enzymes by thermal processing helps to maintain molecular weight.[86]

Moreover, the serum LDL-cholesterol lowering ability of β-glucan from oats was shown in human studies to be dependent on its viscosity. Four meals with different combinations of β-glucan content and molecular weight (4 g of 210 kDa, 3 g of 530 kDa, 4 g of 850 kDa or 3 g of 2210 kDa) were supplied to volunteers for four weeks. All the test meals significantly reduced serum LDL-cholesterol, except the one containing 4 g of the 210 kDa β-glucan.[87] They also demonstrated an inverse relationship between meal viscosity (as measured after *in vitro* digestion) and the reduction in fasting blood cholesterol achieved in the volunteers.[87,88]

Although we are familiar with β-glucan and guar gum being used in food applications, there are novel sources of viscous soluble fibre being investigated around the world. *Detarium senegalense* Gmelin flour is used as a thickening agent in traditional West African cuisine.[89,90] The cell walls from the cotyledon of this legume contain ~60% (w/w) of a water-soluble xyloglucan (molecular weight ~2.7 million). Adding this flour to the normal diet of rats for two weeks, reduced their cholesterol levels and produced no adverse effects. Using the flour in a traditional stew containing 50 g of available carbohydrate, from boiled rice, significantly lowered blood glucose responses in healthy volunteers compared to a low-fibre control. There was also a small, but not statistically significant, reduction in plasma insulin response. When the flour was incorporated into wheat bread the observed reduction in the blood glucose response was thought to be due to the xyloglucan forming a barrier to amylase-starch interactions, in addition to the rheological mechanism, as detailed below for guar GM. The supplemented bread also produced a statistically and physiologically significant attenuation in plasma insulin concentrations compared with the control bread. The reduced plasma glucose and insulin responses were replicated in patients with type 2 diabetes.[89]

3.2 Interactions of NSP with starch and α-amylase

In raw plant foods, starch consisting of amylose and amylopectin, is present in a granular form. The starch granules contain structural regions of ordered and disordered α-glucan chains, with some of the ordered regions comprising crystalline structures. When starch is hydrothermally processed it swells and gelatinises during which the ordered α-glucan structures become more disordered; this process is associated with a dramatic increase in digestibility.[41]

In addition to the viscosity effects produced by guar GM, which are similar to those of oat β-glucan, there is evidence to suggest that it affects the biochemistry of digestion. A wheat bread supplemented with guar GM was fed to pigs fitted with re-entrant cannulas located in the mid-jejunum of the small intestine. Digesta was removed periodically and examined using fluorescently-labelled lectins specific to GM epitopes. Even after 3 hours of digestion, some of the GM seemed to be stuck to the surface of the starch granules, and this interaction was proposed as a potential mechanism by which the GM inhibits α-amylase action on starch during digestion.[91]

However, enzyme kinetic studies by Slaughter and colleagues suggested that the starch-amylase interaction was not affected by the presence of this gum layer, but more likely that guar GM bound to the enzyme, resulting in non-competitive inhibition. These workers also implied that the increased viscosity of the system in the presence of guar did not significantly slow diffusion of maltose, which could potentially act as a competitive inhibitor, in a restricted environment.[92]

Also, there is some evidence to suggest that the presence of soluble NSP restricts the swelling and gelatinisation of starch during hydrothermal processing, which may be part of the explanation of why NSP reduces the rate of amylolysis.[92-94] However, further work is required to confirm the relevance of this mechanism, alongside the rheological and binding mechanisms described above.

3.3 Encapsulation of Nutrients in Intact Cells

The resistance of PCW to digestion may have significant effects on the absorption of nutrients from plant foods and be responsible for some of the effects attributed to DF. Early studies have shown that intact cell walls play a role in hindering starch, lipid and protein digestion and absorption,[95-99] although any physical changes in the cell walls during digestion were not quantified in this early work. The ability of the cell walls to hinder digestion will depend on whether the cells were intact on ingestion, either because the cells separated during mastication, or because relatively large particles remained in the swallowed bolus.

At first glance the chemical and biochemical conditions experienced during digestion seem very severe, and might be expected to degrade polysaccharides easily. However, intact cells of legumes and nuts have been recovered from human ileal effluent and faecal samples,[6,98,101] suggesting they are more stable than the individual components the walls are made from.

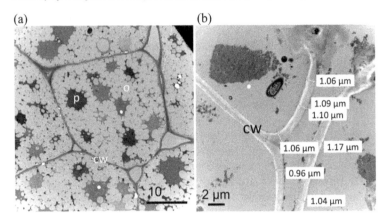

Figure 4 *(a) Raw almond tissue prior to digestion (Scale bar = 10 μm), (b) Almond cell walls after ~20 hours of* in vivo *digestion, showing considerable cell wall swelling (Scale bar = 2 μm). Protein body (p), oleosome (o), cell wall (cw)*

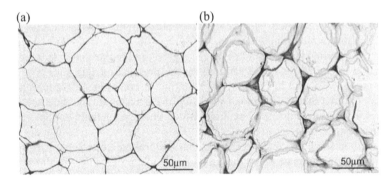

Figure 5 *(a) Raw carrot cell walls prior to digestion, (b) Carrot cell walls after ~10 hours of* in vivo *digestion, showing significant cell wall swelling. Both scale bars = 50μm. Reprinted (adapted) with permission from Tydeman et al.[100]*

Cell walls from different plant sources may have variable degrees of hydration during digestion, which may influence nutrient bioaccessibility. Figure 4 shows that almond cell walls swell by an order of magnitude, from their usual 0.1-0.2 μm, following ~20 h of *in vivo* digestion,[101] but they still appear to be an effective physical barrier. Figure 5 shows the results of a recent study in which carrot cell walls were seen to hydrate to an even greater degree than observed in the almond experiments, and yet still prevented access to the carotene.[100] These studies will be discussed in more detail below.

3.3.1 Almonds (and oily seeds). Almonds (*Amygdalus communis* L.) are nutrient rich dicotyledonous seeds containing, on average 50% lipid, 20% protein, 5% soluble sugar and 12% DF.[102] The almond seed is enclosed in a brown seed coat, which contains ~45% NSP, particularly cellulose and pectin.[103] The majority of the cotyledons is made up of parenchyma cells, which have a relatively thin cell wall (0.1-0.2 μm), and contain multiple small lipid bodies (oleosomes, 1-5 μm) and larger protein inclusions (Figure 4a).[6] Despite their high fat content, almond consumption decreases fasting plasma LDL and oxidised-LDL cholesterol, postprandial glycaemia and insulinaemia, and oxidative damage.[104, 105]

The cells on the surface of almond particles after grinding or chewing are generally fractured, which makes the lipids available for digestion. Figure 6 shows how almond tissue excreted in human faeces was in a relatively intact form apart from the cell walls on the surface of almond particles, which appeared to be susceptible to breakdown/erosion by the microorganisms of the large intestine. It was observed that the particles contained numerous intact parenchyma cells and undigested lipid and protein.[6]

Further evidence that encapsulation by intact cell walls inhibits the rate and extent of lipid digestion in almond seeds was provided by Berry and colleagues.[9] They measured plasma TAG concentrations following meals, consisting of almond muffins with custard, where the almond portion of the meal was presented as almond particles of 2-3 mm or almond oil with defatted almond flour. The peak TAG concentration and the incremental area under the plasma TAG curve were significantly lower for the muffins containing almond particles than for those containing almond flour and oil.

These studies strongly suggest that there is a discrepancy between the predicted calorific value of almonds and the actual energy absorbed. In fact, a recent clinical trial has shown that the energy content of almonds calculated using Atwater factors (6.0 kcal/g) significantly overestimates the actual energy absorbed (4.6 kcal/g).[106] A smaller overestimation (5%) has also been shown for pistachios (*Pistacia vera* L.) using the same methodology.[107]

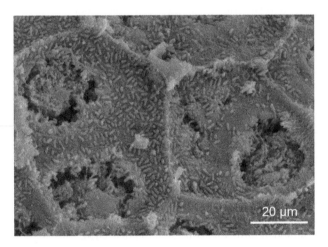

Figure 6 *Breakdown of the cell wall on the surface of an almond tissue particle apparently caused by microflora of the large intestine. Scale bar = 20 μm. Reprinted (adapted) with permission from Ellis et al.*[6]

As the cell walls of almonds act as a barrier to digestive enzymes (at least initially) and almond cells rupture rather than separate during mastication, the development of a predictive model for lipid release from cubic particles of almond was attempted.[108] Application of such a model to chewed almond boluses, where the particle size distributions have been measured using a combination of sieving and laser diffraction has given reasonable predictions of lipid release due to mastication (unpublished data). Current investigations using an *in vitro* model of digestion and ileostomy individuals should provide data on almond cell wall permeability during digestive processes, which it may be possible to incorporate into the predictive model.

3.3.2 Legumes. A number of earlier studies of leguminous seeds have indicated that the cell walls may play an important role in regulating the rate and extent of starch digestion and nutrient bioaccessibility. One of these studies by Noah and colleagues showed that intact cell walls hindered the digestion of starch from white beans (*Phaseolus vulgaris* L.).[95, 98] This result probably explains why legumes produce a relatively low postprandial glycaemic response in human subjects, and are therefore classified as low glycaemic index foods.[11] The cooking of legumes such as kidney beans (*Phaseolus vulgaris* L.) and chickpeas (*Cicer arietinum* L.) leads to a weakening of the middle lamella between cells, facilitating cell separation on mastication. In these foods, the starch is therefore swallowed while it is still seemingly encapsulated by intact cell walls. Therefore, unless the leguminous cells are ruptured, the amylase needs to be able to diffuse through the cell wall to gain access to available starch so that the amylolysis process can progress further.

Penetration of the cell wall by amylase may occur via the plasmodesmata (symplastic transport) or through pores in the cell wall (apoplastic transport). Amylase has a molecular weight of ~56 kDa and a radius of gyration of ~2.7 nm.[109] As plasmodesmata consist of microchannels (radius ~1.25 nm) surrounding a central desmotubule,[110] amylase (and other digestive enzymes) may have to travel through the cell wall matrix itself, which would be dependent on the level of hydration and ionic environment of the wall. Limited information is available on the porosity of leguminous cell walls and the published values are controversial. Baron-Epel and colleagues used fluorescence redistribution after photobleaching (FRAP) with soybean root cells and fluorescently-labelled macromolecules to measure porosity. They reported that macromolecules with Stokes radii ≤3.3 nm could pass through freely, while passage was hindered for molecules with radii in the range 3.3-4.6 nm, and prevented for molecules with radii >4.6 nm.[111] This is larger than the 1.8-2.6 nm determined by Carpita and colleagues using an alternative method in which polyethylene glycols and dextrans of different molecular weight were used to cause cell plasmolysis.[112] The contradictory results may be due to the different methods employed, and also the interpretation of the results. For instance, in the Baron-Epel method, the fluorescently-labelled macromolecules may have remained between the cell wall and the protoplast rather than diffusing into the cytoplasm.[113, 114] Assuming their values apply for leguminous seed porosity, amylase could potentially pass through the walls unhindered, whereas if Carpita's values apply, then amylase would not be able to permeate the cell wall. The low *in vitro* digestibility, and low glycaemic index values reported for legumes, suggest the latter, although further porosity studies are needed.

A second effect to consider in the case of legumes is that the starch is usually tightly packed inside the cell, so that even extended cooking times may not completely gelatinise the starch due to space constraints and/or limited availability of water. If legumes are allowed to cool after cooking, starch digestibility may be reduced due to the formation of retrograded starch, which is known to be less susceptible to amylolysis. *In vitro* studies have shown that retrograded amylose is substantially (>70%) resistant to hydrolysis by amylase.[115] This means leguminous foods may contain a high proportion of resistant starch which is delivered to the colon, where it is fermented, along with the PCW.

3.3.3 Vitamin bioaccessibility. In addition to acting as a barrier to digestion of macronutrients, intact cell walls have been shown to inhibit the bioaccessibility of β-carotene during simulated digestion of carrots (*Daucus carota*).[116] Well-cooked (20 min) carrot cubes released ~2.5% of the carotene present in the gastric phase, whereas raw carrot, of an equivalent size, released ~4.5%. Cooking induced cell separation, but cell wall integrity was maintained. Greater carotene release was produced when carrots were grated (21%) or juiced (97%), due to the increase in the number of ruptured cells. Additionally, significant cell wall swelling during *in vivo* digestion (up to 10 h) did not seem to allow further release of β-carotene from intact cells or induce cell separation, as observed in ileal effluent.[100]

Almonds contain ~27 mg of vitamin E per 100 g, and as such they are said to be a rich source of this vitamin. However, an *in vitro* digestion, simulating gastric and duodenal conditions, showed that 200 μm and 2 mm almond particles released only 45% and 17% of their vitamin E content, respectively.[101]

These results showed that vitamin bioaccessibility is reduced by the presence of intact cell walls in a similar way as for macronutrients, and this information is likely to be important when trying to increase the bioavailability of vitamins from a complex food matrix.

4 CONCLUSIONS

It is clear that the NSP fraction of DF is structurally complex; polysaccharide conformation, composition and interactions with the environment all play a part in how DF behaves during digestion. Nutrition researchers should bear these factors in mind when designing future studies into the health benefits of DF. The presence of NSP as intact cell walls also has an influence on the bioaccessibility of nutrients, including starch, lipid and vitamins. Therefore controlling plant food tissue structure may give food producers/manufacturers ways of optimising their products to produce particular health benefits.

References

1 H. C. Trowell and D. P. Burkitt, *Western Diseases, their Emergence and Prevention*, Harvard University Press, Cambridge, MA, 1981.

2 J. W. Anderson, P. Baird, R. H. Davis, Jr., S. Ferreri, M. Knudtson, A. Koraym, V. Waters and C. L. Williams, *Nutr. Rev.*, 2009, **67**, 188-205.

3 I. A. Brownlee, *Food Hydrocolloid.*, 2011, **25**, 238-250.

4 J. H. Cummings, L. M. Edmond and E. A. Magee, *Clin. Nutr. Suppl.*, 2004, **1**, 5-17.

5 C. W. C. Kendall, A. Esfahani and D. J. A. Jenkins, *Food Hydrocolloid.*, 2010, **24**, 42-48.

6 P. R. Ellis, C. W. C. Kendall, Y. Ren, C. Parker, J. F. Pacy, K. W. Waldron and D. J. A. Jenkins, *Am. J. Clin. Nutr.*, 2004, **80**, 604-613.

7 D. J. A. Jenkins, A. R. Leeds, T. M. S. Wolever, D. V. Goff, K. George, M. M. Alberti, M. A. Gassull, T. Derek and R. Hockaday, *Lancet*, 1976, **308**, 172-174.

8 D. J. A. Jenkins, H. Ghafari, T. M. S. Wolever, R. H. Taylor, A. L. Jenkins, H. M. Barker, H. Fielden and A. C. Bowling, *Diabetologia*, 1982, **22**, 450-455.

9 S. E. E. Berry, E. A. Tydeman, H. B. Lewis, R. Phalora, J. Rosborough, D. Picout and P. R. Ellis, *Am. J. Clin. Nutr.*, 2008, **88**, 922-929.

10 L. Brown, B. Rosner, W. W. Willet and F. M. Sacks, *Am. J. Clin. Nutr.*, 1999, **69**, 30-42.

11 D. J. A. Jenkins, T. M. S. Wolever, A. L. Jenkins, R. G. Josse and G. S. Wong, *Lancet*, 1984, **324**, 388-391.

12 A. R. Bird, M. A. Conlon, C. T. Christophersen and D. L. Topping, *Benef. Microbes*, 2010, **1**, 423-431.

13 M. Elia and J. H. Cummings, *Eur. J. Clin. Nutr.*, 2007, **61 Suppl 1**, S40-74.
14 H. W. Wiley, *Official Methods. Proc. 7th Convention of AOAC*. USDA. Div. of Chemistry. Bulletin No. 25. Page 212, 1890.
15 D. R. Mertens, *J. Anim. Sci.*, 2003, **81**, 3233-3249.
16 E. H. Hipsley, *Br. Med. J.*, 1953, **2**, 420-422.
17 H. Trowell, *Atherosclerosis*, 1972, **16**, 138-140.
18 H. Trowell, D. A. T. Southgate, T. M. S. Wolever, A. R. Leeds, M. A. Gassull and D. J. A. Jenkins, *Lancet*, 1976, **307**, 967.
19 J. W. DeVries, in *Dietary Fibre: new frontiers for food and health*, eds. J. W. van der Kamp, J. Jones, B. V. McCleary and D. L. Topping, Wageningen Academic Publishers, Wageningen, 2010, pp. 29-48.
20 J. I. Mann and J. H. Cummings, *Nutr. Metab. Cardiovas.*, 2009, **19**, 226-229.
21. Codex alimentarius commission, *Report of the 30th session of the codex committee on nutrition and foods for special dietary uses*, ALINORM 09/32/26, www.codexalimentarius.org, 2009.
22 Codex alimentarius commission, *Report of the 31st session of the codex committee on nutrition and foods for special dietary uses*, ALINORM 10/33/26, www.codexalimentarius.org, 2010.
23 J. R. Lupton, in *Dietary Fibre: New Frontiers for Food and Health*, eds. J. W. van der Kamp, J. Jones, B. V. McCleary and D. L. Topping, Wageningen Academic Publishers, Wageningen, 2010, pp. 15-23.
24 P. R. Ellis, Q. Wang, P. Rayment, Y. Ren and S. B. Ross-Murphy, in *Handbook of dietary fibre*, eds. S. S. Cho and M. L. Dreher, Marcel Dekker, Inc., New York, 2001, pp. 613-657.
25 M. Champ, A. M. Langkilde, F. Brouns, B. Kettlitz and Y. L. Bail-Collet, *Nutr. Res. Rev.*, 2003, **16**, 143-161.
26 J. H. Cummings, M. B. Roberfroid, H. Andersson, C. Barth, A. Ferro-Luzzi, Y. Ghoos, M. Gibney, K. Hermonsen, W. P. T. James, O. Korver, D. Lairon, G. Pascal and A. G. S. Voragen, *Eur. J. Clin. Nutr.*, 1996, **51**, 417-423.
27 Y. Ku, O. Jansen, C. J. Oles, E. Z. Lazar and J. I. Rader, *Food Chem.*, 2003, **81**, 125-132.
28 R. G. Crittenden and M. J. Playne, *Trends Food Sci. Technol.*, 1996, **7**, 353-361.
29 K. Swennen, C. M. Courtin and J. A. Delcour, *Crit. Rev. Food Sci. Nutr.*, 2006, **46**, 459-471.
30 B. V. McCleary, J. W. DeVries, J. I. Rader, G. Cohen, L. Prosky, D. C. Mugford, M. Champ and K. Okuma, *J. AOAC Int.*, 2010, **93**, 221-233.
31 B. V. McCleary, J. W. DeVries, J. I. Rader, G. Cohen, L. Prosky, D. C. Mugford, M. Champ and K. Okuma, *J. AOAC Int.*, 2012, **95**, 824-844.
32 F. Kong and R. P. Singh, *J. Food Sci.*, 2008, **73**, R67-R80.
33 S. S. C. Rao, B. Kuo, R. W. McCallum, W. D. Chey, J. D. Dibaise, W. L. Hasler, K. L. Koch, J. M. Lackner, C. Miller, R. Saad, J. R. Semler, M. D. Sitrin, G. E. Wilding and H. P. Parkman, *Clin. Gastroenterol. H.*, 2009, **7**, 537-544.
34 C. Hoebler, A. Karinthi, M. F. Devaux, F. Guillon, D. J. G. Gallant, B. Bouchet, C. Melegari and J. L. Barry, *Br. J. Nutr.*, 1998, **80**, 429-436.
35 M. L. Jalabert-Malbos, A. Mishellany-Dutour, A. Woda and M.-A. Peyron, *Food Qual. Prefer.*, 2007, **18**, 803-812.
36 M.-A. Peyron, A. Mishellany and A. Woda, *J. Dent. Res.*, 2004, **83**, 578-582.
37 C. Brett and K. W. Waldron, *Physiology and Biochemistry of Plant Cell Walls*, Chapman and Hall, London, 1996.

38 C. Hoad, P. Rayment, E. Cox, P. Wright, M. Butler, R. Spiller and P. Gowland, *Food Hydrocolloid.*, 2009, **23**, 833-839.
39 L. Marciani, P. A. Gowland, A. Fillery-Travis, P. Manoj, J. Wright, A. Smith, P. Young, R. Moore and R. C. Spiller, *Am. J. Physiol.-Gastr. L.*, 2001, **280**, G844-G849.
40 S. P. Humphrey and R. T. Williamson, *J. Prosthet. Dent.*, 2001, **85**, 162-169.
41 P. J. Butterworth, F. J. Warren and P. R. Ellis, *Starch-Stärke*, 2011, **63**, 395-405.
42 P. J. Wilde and B. S. Chu, *Adv. Coll. Interfac.*, 2011, **165**, 14-22.
43 R. G. Schipper, E. Silletti and M. H. Vingerhoeds, *Arch. Oral Biol.*, 2007, **52**, 1114-1135.
44 M. Armand, *Curr. Opin. Clin. Nutr.*, 2007, **10**, 156-164.
45 E. Bauer, S. Jakob and R. Mosenthin, *Asian-Austral. J. Anim.*, 2005, **18**, 282-295.
46 D. F. Evans, G. Pye, R. Bramley, A. G. Clark, T. J. Dyson and J. D. Hardcastle, *Gut*, 1988, **29**, 1035-1041.
47 E. Fernández-García, I. Carvajal-Lérida and A. Pérez-Gálvez, *Nutr. Res.*, 2009, **29**, 751-760.
48 G. J. McDougall, I. M. Morrison, D. Stewart and J. R. Hillman, *J. Sci. Food Agric.*, 1996, **70**, 133-150.
49 N. Carpita and D. Gibeaut, *Plant J.*, 1993, **3**, 1-30.
50 B. Buchanan, W. Gruissem and R. Jones, eds., *Biochemistry and molecular biology of plants*, Wiley-Blackwell, USA, 2000.
51 P. J. Harris and B. G. Smith, *Int. J. Food Sci. Technol.*, 2006, **41**, 129-143.
52 J. U. Fangel, P. Ulvskov, J. P. Knox, M. D. Mikkelsen, J. Harholt, Z. A. Popper and W. G. T. Willats, *Front. Plant Sci.*, 2012, **3**, 152.
53 A. J. Parr, K. W. Waldron, A. Ng and M. L. Parker, *J. Sci. Food Agric.*, 1996, **71**, 501-507.
54 A. R. Kirby, A. P. Gunning, K. W. Waldron, V. J. Morris and A. Ng, *Biophys. J.*, 1996, **70**, 1138-1143.
55 M. C. McCann, B. Wells and K. Roberts, *J. Cell Sci.*, 1990, **96**, 323-334.
56 Y. Fang, S. Al-Assaf, G. O. Phillips, K. Nishinari, T. Funami and P. A. Williams, *Struct. Chem.*, 2009, **20**, 317-324.
57 A. Ng and K. W. Waldron, *J. Sci. Food Agric.*, 1997, **73**, 503-512.
58 Y. B. Park and D. J. Cosgrove, *Plant Physiol.*, 2012, **158**, 1933-1943.
59 M. Bunzel, J. Ralph, P. Bruning and H. Steinhart, *J. Agric. Food Chem.*, 2006, **54**, 6409-6418.
60 S. C. Fry, *Phytochem. Rev.*, 2004, **3**, 97-111.
61 M. Bunzel, E. Allerdings, J. Ralph and H. Steinhart, *J. Cereal Sci.*, 2008, **47**, 29-40.
62 T. Ishii, *Carbohydr. Res.*, 1991, **219**, 15-22.
63 S. V. Levigne, M.-C. J. Ralet, B.-C. L. Quéméner, B. N. Pollet, C. Lapierre and J.-F. J. Thibault, *Plant Physiol.*, 2004, **134**, 1173-1180.
64 M. Bunzel, J. Ralph, J. M. Marita, R. D. Hatfield and H. Steinhart, *J. Sci. Food Agric.*, 2001, **81**, 653-660.
65 C. Funk, A. Braune, J. H. Grabber, H. Steinhart and M. Bunzel, *J. Agric. Food Chem.*, 2007, **55**, 2418-2423.
66 P. J. Harris and J. A. K. Trethewey, *Phytochem. Rev.*, 2010, **9**, 19-33.
67 D. T. A. Lamport, M. J. Kieliszewski, Y. Chen and M. C. Cannon, *Plant Physiol.*, 2011, **156**, 11-19.
68 D. J. Cosgrove, *Nat. Rev. Mol. Cell Biol.*, 2005, **6**, 850-861.
69 L. Goulao and C. Oliveira, *Trends Food Sci. Technol.*, 2008, **19**, 4-25.
70 W. B. Herppich and S. Huyskens-Keil, *Ann. Appl. Biol.*, 2008, **152**, 377-388.

71 E. Jamet, C. Albenne, G. Boudart, M. Irshad, H. Canut and R. Pont-Lezica, *Proteomics*, 2008, **8**, 893-908.
72 H. Meier and J. S. G. Reid, in *Plant Carbohydrates I: Intracellular carbohydrates*, eds. L. F.A. and W. Tanner, Springer-Verlag, Berlin, 1982, vol. 13A, pp. 418-471.
73 M. L. Parker, *Protoplasma*, 1984, **120**, 233-241.
74 K. Nishinari, M. Takemasa, H. Zhang and R. Takahashi, in *Comprehensive Glycoscience*, ed. J. P. Kamerling, Elsevier B.V., 2007, vol. 2.
75 S. C. Chuang, T. Norat, N. Murphy, A. Olsen, A. Tjønneland, K. Overvad, M. C. Boutron-Ruault, F. Perquier, L. Dartois, R. Kaaks, B. Teucher, M. M. Bergmann, H. Boeing, A. Trichopoulou, P. Lagiou, D. Trichopoulos, S. Grioni, C. Sacerdote, S. Panico, D. Palli, R. Tumino, P. H. Peeters, B. Bueno-de-Mesquita, M. M. Ros, M. Brustad, L. A. Åsli, G. Skeie, J. R. Quirós, C. A. González, M.-J. Sánchez, C. Navarro, E. Ardanaz Aicua, M. Dorronsoro, I. Drake, E. Sonestedt, I. Johansson, G. Hallmans, T. Key, F. Crowe, K.-T. Khaw, N. Wareham, P. Ferrari, N. Slimani, I. Romieu, V. Gallo, E. Riboli and P. Vineis, *Am. J. Clin. Nutr.*, 2012, **96**, 164-174.
76 Y. Park, A. F. Subar, A. Hollenbeck and A. Schatzkin, *Arch. Intern. Med.*, 2011, **171**, 1061-1068.
77 P. Rayment, S. B. Ross-Murphy and P. R. Ellis, *Carbohyd. Polym.*, 1995, **28**, 121-130.
78 P. Rayment, S. B. Ross-Murphy and P. R. Ellis, *Carbohyd. Polym.*, 1998, **35**, 55-63.
79 C. L. Dikeman and G. C. Fahey, *Crit. Rev. Food Sci. Nutr.*, 2006, **46**, 649-663.
80 P. R. Ellis, F. G. Roberts, A. G. Low and L. M. Morgan, *Br. J. Nutr.*, 1995, **74**, 539-556.
81 D. Oakenfull, in *Handbook of dietary fibre*, eds. S. S. Cho and M. L. Dreher, Marcel Dekker Inc, New York, 2001, pp. 195-206.
82 Q. Wang, P. R. Ellis and S. B. Ross-Murphy, *Carbohyd. Polym.*, 2006, **64**, 239-246.
83 Q. Wang, P. R. Ellis and S. B. Ross-Murphy, *Carbohyd. Polym.*, 2008, **74**, 519-526.
84 X. Lan-Pidhainy, Y. Brummer, S. M. Tosh, T. M. S. Wolever and P. J. Wood, *Cereal Chem.*, 2007, **84**, 512-517.
85 S. M. Tosh, Y. Brummer, T. M. S. Wolever and P. J. Wood, *Cereal Chem.*, 2008, **85**, 211-217.
86 A. Regand, S. M. Tosh, T. M. S. Wolever and P. J. Wood, *J. Agric. Food Chem.*, 2009, **57**, 8831-8838.
87 T. M. S. Wolever, S. M. Tosh, A. L. Gibbs, J. Brand-Miller, A. M. Duncan, V. Hart, B. Lamarche, B. A. Thomson, R. Duss and P. J. Wood, *Am. J. Clin. Nutr.*, 2010, **92**, 723-732.
88 S. M. Tosh, Y. Brummer, S. S. Miller, A. Regand, C. Defelice, R. Duss, T. M. S. Wolever and P. J. Wood, *J. Agric. Food Chem.*, 2010, **58**, 7723-7730.
89 P. A. Judd and P. R. Ellis, in *Traditional Medicines for Modern Times*, ed. A. Soumyanath, 2005.
90 U. A. Onyechi, P. A. Judd and P. R. Ellis, *Br. J. Nutr.*, 1998, **80**, 419-428.
91 C. S. Brennan, D. E. Blake, P. R. Ellis and J. D. Schofield, *J. Cer. Sci.*, 1996, **24**, 151-160.
92 S. L. Slaughter, P. R. Ellis, E. C. Jackson and P. J. Butterworth, *Biochim. Biophys. Acta*, 2002, **1571**, 55-63.
93 S. L. Slaughter, P. R. Ellis and P. J. Butterworth, *Biochim. Biophys. Acta*, 2001, **1525**, 29-36.
94 R. F. Tester and M. D. Sommerville, *Food Hydrocolloid.*, 2003, **17**, 41-54.
95 A. Golay, A. M. Coulston, C. B. Hollenbeck, L. L. Kaiser, P. Würsch and G. M. Reaven, *Diabetes Care*, 1986, **9**, 260-266.
96 A. S. Levine and S. E. Silvis, *N. Engl. J. Med.*, 1980, **303**, 917-918.

97   C. Melito and J. Tovar, *Food Chem.*, 1995, **53**, 305-307.
98   L. Noah, F. Guillon, B. Bouchet, A. Buléon, C. Molis, M. Gratas and M. Champ, *J. Nutr.*, 1998, **128**, 977-985.
99   P. Würsch, S. Del Vedovo and B. Koellreutter, *Am. J. Clin. Nutr.*, 1986, **43**, 25-29.
100 E. A. Tydeman, M. L. Parker, R. M. Faulks, K. L. Cross, A. Fillery-Travis, M. J. Gidley, G. T. Rich and K. W. Waldron, *J. Agric. Food Chem.*, 2010, **58**, 9855-9860.
101 G. Mandalari, R. Faulks, G. T. Rich, V. Lo Turco, D. Picout, R. B. Lo Curto, C. Bisignano, G. Dugo, K. W. Waldron, P. R. Ellis and M. S. J. Wickham, *J. Agric. Food Chem.*, 2008, **56**, 3409-3416.
102 S. Yada, K. G. Lapsley and G. Huang, *J. Food Compos. Anal.*, 2011, **24**, 469-480.
103 G. Mandalari, A. Tomaino, T. Arcoraci, M. Martorana, V. Lo Turco, F. Cacciola, G. T. Rich, C. Bisignano, A. Saija, P. Dugo, K. L. Cross, M. L. Parker, K. W. Waldron and M. S. J. Wickham, *J. Food Compos. Anal.*, 2010, **23**, 166-174.
104 D. J. A. Jenkins, C. W. C. Kendall, D. A. Faulkner, T. Nguyen, T. Kemp, A. Marchie, J. M. W. Wong, R. de Souza, A. Emam, E. Vidgen, E. A. Trautwein, K. G. Lapsley, C. Holmes, R. G. Josse, L. A. Leiter, P. W. Connelly and W. Singer, *Am. J. Clin. Nutr.*, 2006, **83**, 582-591.
105 A. R. Josse, C. W. C. Kendall, L. S. Augustin, P. R. Ellis and D. J. A. Jenkins, *Metabolism*, 2007, **56**, 400-404.
106 J. A. Novotny, S. K. Gebauer and D. J. Baer, *Am. J. Clin. Nutr.*, 2012, **96**, 296-301.
107 D. J. Baer, S. K. Gebauer and J. A. Novotny, *Br. J. Nutr.*, 2012, **107**, 120-125.
108 P. R. Ellis, D. R. Picout, M. S. J. Wickham, G. Mandalari, G. T. Rich, R. M. Faulks and K. G. Lapsley, *FASEB J.*, 2007, **21**, A119.
109 I. Simon, S. Móra and P. Elödi, *Mol. Cell. Biochem.*, 1974, **4**, 211-216.
110 X. Wu, D. Weigel and P. A. Wigge, *Gene. Dev.*, 2002, **16**, 151-158.
111 O. Baron-Epel, P. K. Gharyal and M. Schindler, *Planta*, 1988, **175**, 389-395.
112 N. C. Carpita, D. Sabularse, D. Montezinos and D. P. Delmer, *Science*, 1979, **205**, 1144-1147.
113 N. C. Carpita, *Science*, 1982, **218**, 813-814.
114 M. Tepfer and I. E. P. Taylor, *Science*, 1981, **213**, 761-763.
115 R. L. Botham, V. J. Morris, T. R. Noel and S. G. Ring, *Carbohyd. Polym.*, 1996, **29**, 347-352.
116 E. A. Tydeman, M. L. Parker, M. S. J. Wickham, G. T. Rich, R. M. Faulks, M. J. Gidley, A. Fillery-Travis and K. W. Waldron, *J. Agric. Food Chem.*, 2010, **58**, 9847-9854.

5

STABILITY AND DEGRADATION PATHWAYS OF POLYSACCHARIDE AND GLYCOCONJUGATE VACCINES

C. Jones

Laboratory for Molecular Structure, NIBSC, Blanche Lane, South Mimms, Herts EN6 3QG, UK
Chris.Jones@nibsc.hpa.org.uk

1 INTRODUCTION

Encapsulated bacteria, expressing a polysaccharide capsule, remain major causes of mortality and morbidity, especially amongst the young in Developing Countries. The pathogens include *Haemophilus influenzae*, *Streptococcus pneumoniae*, *Salmonella enterica* serovar Typhi, and *Neiserria meningitidis*. Together, these organisms cause up to 2 million child deaths per year. The capsule (Figure 1) acts to prevent dehydration on the bacterium and to modulate the flow of nutrients to the cell. When an infection is established, it hides cell surface components from the host's innate immune systems.

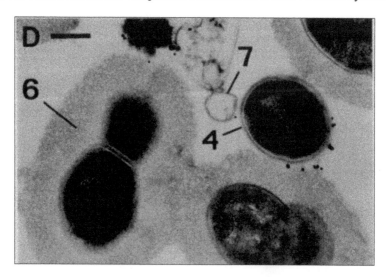

Figure 1 *Electron micrograph of* Streptococcus pneumoniae *stained in such a way as to prevent collapse of the polysaccharide capsule (marked as 6). The cell marked as 4 is unencapsulated. Reproduced from Skov Sørensen et al., Infect. Immun. 1988, 56, 1890-1896 with permission*

Table 1 *Structures of repeat units of some vaccine-relevant bacterial polysaccharides*

Bacterium	Repeat unit stricture
H. influenzae Type b	→3DRib*f*/β1→1DRib'ol5→OPO$_3$→
N. meningitidis Group A	→6DManNAc(3OAc)α1→OPO$_3$→
N. meningitidis Group C	→9DNeu5Ac(7/8OAc)α2→
N. meningitidis Group W135	→6DGalα1→4DNeu5Ac(9OAc)α2→
N. meningitidis Group Y	→6DGlcα1→4DNeu5Ac(9OAc)α2→
S. Typhi	→4DGalNAcA(3OAc)α1→
S. pneumoniae Type 4	→3DManNAcβ1→3LFucNAcα1→3DGalNAcα1→ 4DGal(2,3SPyr)α1→
S. pneumoniae Type 14	→6DGlcNAcβ1→3DGalβ1→4DGlcβ1→ 4 | βDGal

A single pathogenic species may contain many strains expressing different capsular polysaccharides – these are usually called serotypes or serogroups depending on the organism. Of the six serotypes of *H. influenzae* serotype b (called Hib) causes the large majority of disease and a monovalent vaccine is sufficient. On the other hand there are more than 90 serotypes of *S. pneumoniae* and many cause disease, so that vaccines need to contain multiple polysaccharides.[1,2] In *S. pneumoniae* and *N. menigitidis* the major disease-causing serotypes/ serogroups vary geographically, with patient age and temporally and different vaccines may be optimal for different regions. For example, meningococcal serogroup A, generally rare elsewhere, causes epidemics in sub-Saharan Africa and a dedicated vaccine has been developed.[3] Of the twelve meningococcal serogroups five have historically been associated with disease, and some favour different age groups, although a sixth serogroup, serogroup X, is currently emerging as a significant problem in Africa.[4] Structurally CPSs are repeating polymers, either linear or branched. They have a strict repeat unit which can be a single monosaccharide or an oligosaccharide of up to about eight monosaccharide units.[1,2] Rare monosaccharide residues found only in bacteria may be present.[5] Non-carbohydrate substituents, including *O*-acetyl groups and phosphorylated derivatives, may be also present. Incomplete *O*-acetylation or *O*-acetyl group migration between hydroxyl groups is the only significant source of heterogeneity.[6] Structure determination of the repeat units has been an active area of research for many years.[7-9] The structures of some of the key polysaccharides are shown in Table 1.

2 PURIFIED POLYSACCHARIDE VACCINES

The first generation of polysaccharide-based vaccines are either single purified polysaccharides (the typhoid Vi polysaccharide antigen) or blends of four (meningococcal) or twenty three (pneumococcal vaccines), and contain either 25µg (Vi and pneumococcal) or 50 µg (meningococcal vaccines) of each serotype-specific polysaccharide. Polysaccharides are T cell-independent Type II immunogens and elicit an antibody response by an immunological pathway that only develops in the second year of life, so that these vaccine not able to create protective immune responses in young infants.[1,2] The vaccines are designed for use in adolescents and adults. The standard model for the molecular mechanism of action, (summarised in Figure 2) involves crosslinking of cell surface receptors, membrane-bound immunoglublins or mIg, (between 10 and 20) on the B cell, leading through a series of protein phosphorylation steps to an increase in internal free calcium.[10] This primed cell is activated to differentiate and produce antibodies by a second signal, the nature of which is less well understood. These vaccines require only a single dose to elicit a protective immune response in adults, but because immunological memory is not invoked regular revaccination (typically at 5 year intervals) is required.

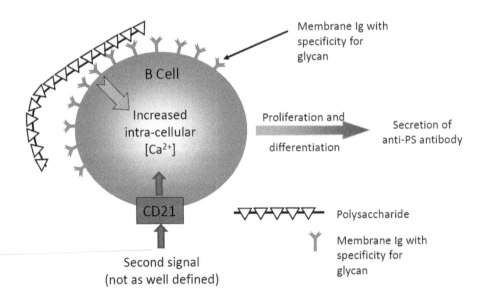

Figure 2 *Model of the mechanism by which T cell independent Type II immunogens, such as repeating polysaccharides induce an immune response.[10] The high molecular weight polysaccharide crosslinks surface expressed immunoglobulin present on B cell, leading, through a series of protein phosphorylation steps to an increase in free intracellular calcium. A second signal, less well defined, then causes these primed B cells to differentiate and secrete antibody*

Critical quality attributes[11] for polysaccharides used in vaccine products are (i) identity, which is defined by the structure of the polysaccharide repeat unit, (ii) molecular size, as low molecular weight polysaccharides are unable to crosslink sufficient B cell surface receptors and so are not immunogenic, and (iii) safety factors such as endotoxin content.

Identity is confirmed by immunological methods, such as ELISA, analysis of composition or one-dimensional ^{1}H NMR methods as the repeat units have distinctive profiles. Because of the strict repeat units, NMR spectra are relatively simple, and T_2 relaxation is dominated by internal mobility, leading to much higher resolution spectra than would be expected for high molecular weight polymers.[12] NMR methods easily detect minor structural variations, such as the differences in linkage position, and can be used to quantify labile substituents such as *O*-acetyl groups. This involves either visual or mathematical[13] comparison of test and reference spectra, or by *in situ* base-catalysed de-*O*-acetylation and relative quantification of the derived acetate anion.[12] The importance of *O*-acetylation in defining immunogenicity, and a protective response in particular, remains a topic of debate and probably differs between polysaccharides.

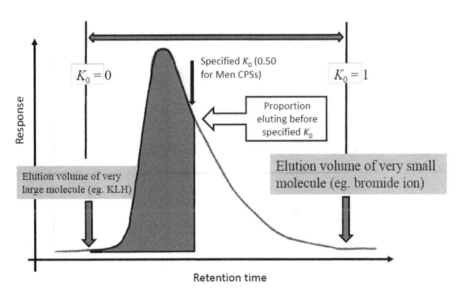

Figure 3 *Specifications used in the European Pharmacopoeia for the molecular sizing of bacterial polysaccharides used in typhoid and meningococcal vaccines, in which a specified proportion of the polysaccharide much elute before a defined K_D (distribution coefficient) value on a column containing a defined matrix. In contrast, for pneumococcal polysaccharides, the peak of the curve must elute before a defined K_D on a column containing a defined matrix*

Conventionally, molecular sizing is achieved by soft gel permeation chromatographic on defined column matrices,[11] with specification either relating to the K_D (distribution coefficient) of the peak or the proportion of material eluting before a defined K_D (Figure 3). Detection is typically by refractive index or, if analysing polyvalent vaccines, by serotype-specific immunological methods such as rate nephelometry. These methods are being replaced in the vaccine industry by high performance size exclusion chromatographic methods with multiple angle laser light scattering (MALS) detection. When the refractive index increment, dn/dc, is known, this allows molecular size specifications to be correlated with molecular weight.[14] The smallest polysaccharides used in vaccine manufacture, typically meningococcals, have molecular weights in excess of 100kDa, whilst pneumococcal polysaccharides often exceed 1 MDa. Degradation of the polysaccharide which leads to a reduction in molecular size (or molecular weight) is expected to lead to a reduction in immunogenicity and vaccine efficacy.

Since only high molecular weight polysaccharides are immunogenic, physicochemical methods such as NMR for detecting novel endgroups formed during depolymerisation, are rarely sufficiently sensitive for pharmaceutical analysis. Immunochemical detection methods are not sensitive to variation in polysaccharide molecular weight.

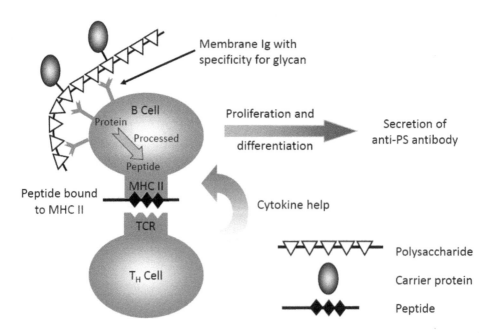

Figure 4 *Model for the induction of an immune response by a glycoconjugate vaccine. The conjugate binds to an antigen-presenting cell (APC) presenting receptors specific for the saccharide chain. The glycoconjugate is internalised and the carrier protein subjected to proteolytic digestion. Carrier-derived peptides are displayed on MHC Type II molecules, which interact with T cells. This leads to direct cell-cell interactions and indirect interactions through cytokine secretion which leads to APC differentiation and antibody secretion*[15]

3 GLYCOCONJUGATE VACCINES

As the major risk group for diseases caused by encapsulated bacteria are young infants, there was a strong imperative to develop similar vaccines against these pathogens which are effective in infants.[1,2] This is achieved by covalent attachment of the polysaccharide, or an oligosaccharide derived from the polysaccharide, to a suitable carrier protein. These glycoconjugate vaccines use a different molecular mechanism to invoke an immune response, with T cell participation, that has already developed in infants.

The standard model[15] is illustrated in Figure 4, although this is undoubtedly an oversimplification. T cell participation also results in avidity maturation, antibody isotype switching and induction of immunological memory – overall a more effective profile of antibodies for protection against infection.

Because a different molecular mechanism is used the dependence of immunogenicity on molecular weight is very different, so that vaccines can be manufactured using oligosaccharide haptens (and in future vaccines against a wider variety of pathogens may be possible, provided that the anti-saccharide antibodies protect against disease). For these vaccines several immunisations are required before infants develop a protective immune response, and with a booster in the second year to ensure long-lived protection. Other immunisation regimes may also prove appropriate.

Existing glycoconjugate vaccines[1,2] can be grouped into three basic structural models shown in Figure 5. Activation of the polysaccharide with depolymerisation (and 10-20 repeat units is a common specification for the sized oligosaccharides) and attachment to a carrier protein in such a way that there is limited crosslinking of the carrier proteins leads to neoglycoprotein immunogens (Figure 5A). With CRM197, a genetically toxoided diphtheria toxin, as the carrier and a polysaccharide protein ratio of ca. 0.35 this gives rise to a molecule of approximately 90kDa. These molecules, with short highly flexible glycan chains can be readily characterised by NMR, optical spectroscopy and chromatographic approaches. Random multisite activation of intact high molecular weight polysaccharide chains, without depolymerisation, and coupling to a carrier protein (typically tetanus toxoid, diphtheria toxoid or CRM197) leads to high mass crosslinked network immunogens: a mass of 5 MDa has been cited for Hib-TTx conjugates (Figure 5B). These immunogens are less easy to characterise by physicochemical methods. Finally, attachment of "size reduced" polysaccharide chains to LPS-depleted outer membrane protein (OMP) vesicles gives rise to a third structural class (Figure 5C): these vaccines invoke immune responses different to the two previously described classes.

Polysaccharides do not react with potential carrier proteins without chemical activation, and the optimal conjugation chemistry depends principally on the structure of the CPS repeat unit, and reactive groups within that.[1,2] Polysaccharides contain (or can be modified to contain) a number of groups which can be used in activation, including acid-labile glycocosidic linkages, *cis*-diols, carboxylate groups in uronic acids, unsubstituted amino groups, and phosphomonoesters. In addition, reagents such as cyanogen bromide and carbonyldiimidazole react with hydroxyl groups in the polysaccharide, allowing a random activation to which a linker can be attached. In some cases the carrier protein is also chemically activated to provide complementary reactivity for polysaccharide conjugation.

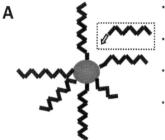

A

- Produced by coupling monofunctional oligosaccharides
- Produced by coupling bifunctional oligosaccharide at low coupling efficiencies.
- Either direct or indirect attachment to carrier protein (ie. through linker)
- Most often used with CRM197 as carrier, producing a conjugate with MW ca. 90kDa and 30% w/w carbohydrate
- Similar to a typical plasma protein

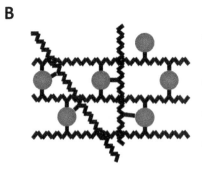

B

- Produced by random activation of high mass polysaccharides, with multiple activations per chain.
- Coupled to carrier protein through non-specific chemistry
- Each polysaccharide chain attached to multiple carrier proteins
- Each carrier protein coupled to multiple polysaccharide chains
- Often used with tetanus toxoid as carrier protein
- Network of high mass (typically 5MDa for a Hib conjugate)

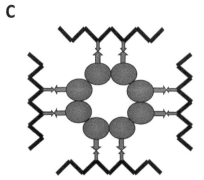

C

- Produced by random activation of reduced mass polysaccharides, with multiple activations per chain.
- "Carrier protein" is an LPS-depleted mixture of outer membrane proteins
- Vesicle nature of OMPs creates a high mass complex.
- OMPs were chosen to provide complementary immunological detection.
- Hard to make materials on a very large scale.

Figure 5 *Cartoons depicting the three basic structural types of glycoconjugate vaccine in current use. A. Glycoconjugates produced from oligosaccharides with a single attachment site, or limited crosslinking through two sites. B. Glycoconjugates produced by multiple activation sites on high mass polysaccharides and multiple attachments to the carrier protein, producing crosslinked network, and C. Glycoconjugates produced by attachment of "size-reduced" polysaccharides to outer membrane protein vesicles through multiple attachment sites*

4 CONJUGATE VACCINES USING OLIGOSACCHARIDES

Concomitant activation and depolymerisation of some polysaccharides, including the vaccine relevant Hib and meningococcal Group C polysaccharides can be achieved either by controlled dilute acid hydrolysis (targeting the ribofuranose or sialic acid residues respectively, which have acid-labile glycosidic linkages), or by periodate oxidation, reacting with the *cis*-diols present in both and cleaving the C-C bond. The periodate oxidation of Hib CPS is shown in Figure 6, and the newly created aldehydic groups can be used to attach the oligosaccharides to the carrier protein through reductive amination reactions with the ε-amino groups and N-terminal amino group of the carrier protein. The mean molecular size of the oligosaccharide can be determined by NMR spectroscopy, comparing the intensity of the resonances from newly created end groups with those of intact repeat units, of the profile assesses by anion exchange chromatography.[16,17] CRM197 has been widely used as a carrier for vaccines of this type. The product is a neoglycoconjugate vaccine of the type shown in Figure 5A.

In other cases, such as the meningococcal Groups A, W135 and Y CPSs and the pneumococcal polysaccharides in heptavalent Prevenar, a pneumococcal glycoconjugate vaccine, controlled periodate oxidation cleaves *cis*-diols without chain depolymerisation, so that reductive amination conjugation chemistry attaches the carrier protein to, and crosslinks, high molecular mass polysaccharide chains. CRM197, tetanus toxoid and *Haemophilus* Protein D have been used as carriers in vaccines of this type, and the product is a crosslinked vaccine of the type depicted in Figure 5B. Most other conjugation chemistries retain the high mass of the original polysaccharide.

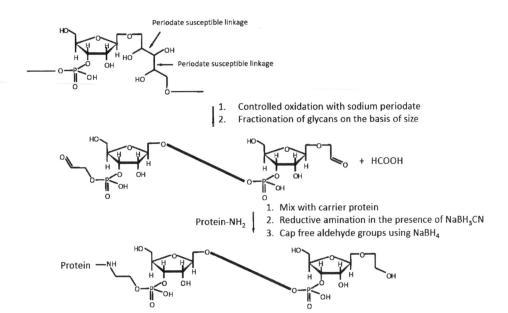

Figure 6 *Schematic showing the controlled periodate cleavage of the* Haemophilus influenzae *type b polysaccharide and conjugation to free amino groups present on the carrier protein. Residual aldehydic groups remaining on the polysaccharide after conjugation are capped by reduction*

5 STABILITY OF GLYCOCONJUGATE VACCINES

Ensuring the structural integrity of vaccines throughout their shelf life and until they are delivered to the patient is a critical aspect of any vaccine quality programme. Glycoconjugate vaccine instability could arise, from amongst other factors, (i) denaturation or degradation of the carrier protein, (ii) by separation of the intact saccharide chain from the carrier protein (deconjugation), or (iii) from degradation and depolymerisation of the attached glycan chains. The first of these is outside the remit of this chapter, the second depends on the choice of conjugation chemistry and the last is the subject of this section. The earliest conjugate vaccine, against Hib and Group C meningococcal infections, happened to use polysaccharide chains of unusually low stability, and the major cause of degradation arises from depolymerisation of the chains. Mechanisms of glycan depolymerisation can be investigated by NMR spectroscopy.

The Hib polysaccharide degrades by rearrangement of the phosphodiester linkage, initially forming one of two cyclic phosphodiesters (Figure 7) with chain cleavage. These can hydrolyse, rapidly under basic conditions, to two possible phosphomonoesters, with the majority (ca. 90%) of the chains retaining the phosphate group on the ribose, rather than the ribitol, residue. Deliberate base hydrolysis to a mixture of nine phosphorylated variants of the repeat unit, separated by high performance anion exchange chromatography, is used analytically to quantify Hib polysaccharide in vaccines.

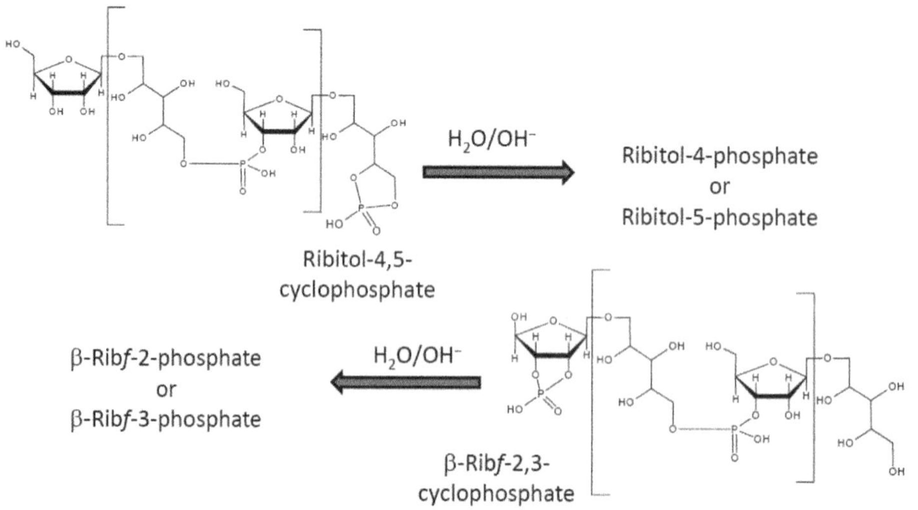

Figure 7 *Initial cyclophosphodiester degradation products of the* Haemophilus influenzae *type b (Hib) polysaccharide, which can undergo hydrolysis to phosphomonoesters*

The novel phosphorylated end groups have characteristic NMR signatures, and can be identified in ^{1}H-^{13}C HSQC spectra of deliberately degraded material (Figure 8), and full spectral assignments obtained by homo- and hetero-nuclear correlation experiments. These fingerprint resonances can also be observed[15] in 500MHz 1D ^{1}H NMR spectra of deliberately degraded Hib-CRM197 bulk vaccine (Figure 9).

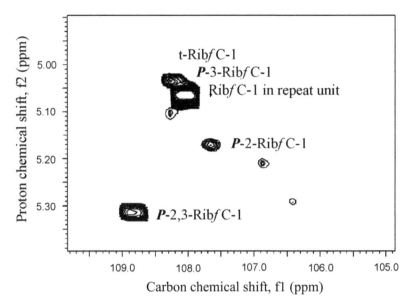

Figure 8 *Partial 500MHz ^{1}H-^{13}C HSQC spectrum of the anomeric region of thermally degraded (several days at room temperature) Hib polysaccharide. Anomeric resonances are labelled from the phosphorylation pattern of the ribofuranose residue*

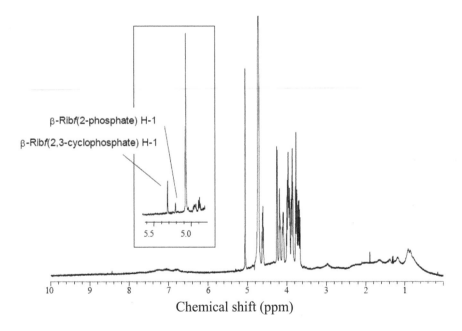

Figure 9 *Partial 500MHz 1D ^{1}H spectrum of intact (main spectrum) and deliberately degraded (inset) Hib-CRM197 bulk conjugate vaccine. The additional resonances arise from the anomeric hydrogens of novel end groups formed during the degradation of the Hib polysaccharide chains*

The Men C polysaccharide tends to degrade by hydrolysis of the glycosidic linkage (Figure 10), leading to (i) migration of the O-acetyl group present at either O-8 or O-7 to the thermodynamically favoured O-9, and (ii) equilibriation at the now-unsubstituted anomeric position to form, mainly the β-anomer. The rearrangement of the O-acetylation pattern leads to resolved downfield H-9 and H-9' resonances, whilst the H-3a/H-3e crosspeak from the β-anomer of sialic acid is visible[19] in TOCSY spectra (Figure 10). Whilst these NMR approaches are extremely valuable for detailed studies of mechanism they are too complex for routine assays, where GPC methods or measurement of free (or unconjugated) saccharide are preferred.

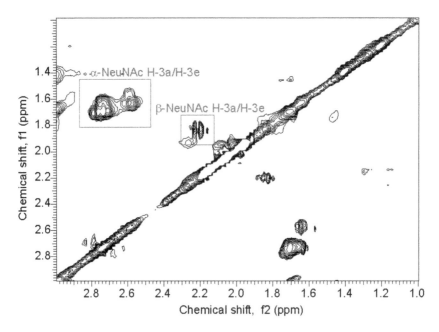

Figure 10 *Partial 500MHz 80msec TOCSY spectrum of a CRM197-MenC conjugate vaccine, showing the H-3a/H-3e crosspeak which arises from the β-anomer of sialic acid residues: this anomer is only possible due to equilibriation after hydrolysis of the MenC polysaccharide chain. Reproduced with the permission of Karger, Basel, from ref. 19*

6 SUMMARY

Overall, successful development of polysaccharide-based vaccines depends upon an understanding of the mechanisms for controlled and uncontrolled degradation of the glycan chains, firstly in the development of conjugation chemistries matched to the structure of the polysaccharide repeat units, and secondly to ensure the continuing integrity of the product in stability studies. A range of analytical techniques are available to characterise these products, with methods such as size exclusion chromatography, often linked to MALS detection, especially powerful for assessing molecular size, and NMR spectroscopy to investigate the molecular mechanisms of degradation.

References

1 C. Jones, *An. Bras. Acad. Cienc.,* 2005, **77**, 293; available at http://www.scielo.br/pdf/aabc/v77n2/a09v77n2.pdf

2 C. Jones, in *Comprehensive Glycoscience,* (J.P. Kamerling, Ed.), Elsevier, Amsterdam, 2007, p. 569.

3 C. Frasch, M.P. Preziosi, and F.M. Laforce, *Hum. Vaccin. Immunother.*, 2012, **8**, 1.

4 I. Delrieu, S. Yaro, T.A. Tamekloé, B.M. Njanpop-Lafourcade, H. Tall, P. Jaillard, M.S. Ouedraogo, K. Badziklou, O. Sanou, A. Drabo, B.D. Gessner, J.L. Kambou and J.E. Mueller, *PLoS One,* 2011, **6**, e19513.

5 B. Lindberg, *Adv. Carbohydr. Chem. Biochem.*, 1990, **48**, 279.

6 X. Lemercinier and C. Jones, *Carbohydr. Res.*, 1996, **296**, 83.

7 C. Jones and X. Lemercinier, *Carbohydr. Res.*, 2005, **340**, 403.

8 C. Jones, *Carbohydr. Res.*, 1995, **265**, 175.

9 C. Jones, *Carbohydr. Res.,* 2005, **340**, 1097.

10 Q. Vos, A. Lees, Z.Q. Wu, C.M. Snapper and J.J. Mond, *Immunol. Rev.*, 2000, **176**, 154.

11 *European Pharmacopoeia* version 7.5, European Pharmacopoeia, Strasbourg, 2012, p. 803, 813 and 836.

12 C. Jones, *J. Pharm. Biomed. Anal.*, 2005, **38**, 840.

13 C. Abeygunawardana, T.C. Williams, J.S. Sumner and J.P. Hennessey Jr., *Anal. Biochem.,* 2000, **279**, 226.

14 J.E. MacNair, T. Desai, J. Teyral, C. Abeygunawardana and J.P. Hennessey Jr., *Biologicals*, 2005, **33**, 49.

15 G.R. Siber, *Science,* 1994, **265**, 1385.

16 N. Ravenscroft, G. Averani, A. Bartoloni, S. Berti, M. Bigio, V. Carinci, P. Costantino, S. D'Ascenzi, A. Giannozzi, F. Norelli, C. Pennatini, D. Proietti, C. Ceccarini and P. Cescutti, *Vaccine,* 1999, **17**, 2802

17 P. Costantino, F. Norelli, A. Giannozzi, S. D'Ascenzi, A. Bartoloni, S. Kaur, D. Tang, R. Seid, S. Viti, R. Paffetti, M. Bigio, C. Pennatini, G. Averani, V. Guarnieri, E. Gallo, N. Ravenscroft, C. Lazzeroni, R. Rappuoli and C. Ceccarini, *Vaccine,* 1999, **17**, 1251.

18 C. Jones, D.T. Crane, X. Lemercinier, B. Bolgiano and S.E. Yost, *Dev. Biol. Stand.,* 1996, **87**, 143.

19 C. Jones, X. Lemercinier, D.T. Crane, C.K. Gee and S. Austin, *Dev. Biol. (Basel)*, 2000, **103**, 121.

6
MUCIN TURNOVER

A.P. Corfield

Mucin Research Group, School of Clinical Sciences, University of Bristol, Bristol Royal Infirmary, Marlborough Street, Bristol, BS2 8HW, UK
corfielda@gmail.com

1 INTRODUCTION

Protection at mucosal surfaces throughout the body have developed to maintain a constant and dynamic screen of the external environment. Many different processes are necessary to allow protection against all manner of agents which may cause damage, while enabling vital interactions required for normal survival. Recovery and transport of nutrients from the diet must be balanced against resistance to harmful chemicals, toxins and other potentially damaging molecules. At the same time, a constant communication with the microflora is needed to utilise those beneficial commensal strains while resisting pathogens. Provision of a milieu enabling the action of a range of antimicrobial molecules is also required. Thus, the mucosal protective barrier must be a dynamic system giving continuous defence while enabling normal and life supporting processes. Biomarkers for mucosal disease reflect the failure of the protective barrier at a variety of levels and underline its' multifactorial nature.

The mucins are a fundamental component found in the protective mucosal barrier. They have a novel molecular structure allowing the formation of viscoelastic gels, thus donating a molecular network which incorporates the mucosal surface factors described above. The barrier is closely integrated with both innate and immune defensive systems and must be turned over in a controlled manner in order to maintain defensive functions on a continuous basis[1-5]. This chapter is largely focussed on the mucosal protective barrier in the human gastrointestinal tract (GI). Many of the features described here are mimicked at other mucosal surfaces throughout the human body, where tissue specific adaptations are found to tally with particular demands.

2 STRUCTURE OF THE MUCOSAL BARRIER

The gastrointestinal mucosal barrier is composed of a secreted adherent mucus gel layer, having strongly and weakly (sloppy) adherent components. The strongly adherent layer is in contact with the glycocalyx, anchored to mucosal cell-surface membranes[2-10]. The thickness and composition of the mucus layer varies throughout the GI tract. The small intestine shows the thinnest layer, 150 to 300 µm, while both the stomach and colon have

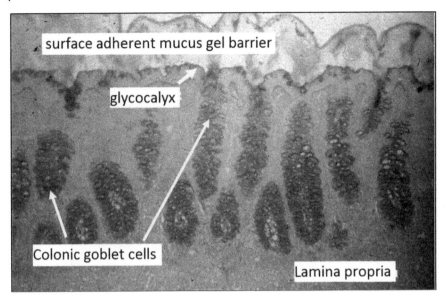

Figure 1 *Mucin layers in the human colorectum. The mucosal barrier is shown including the adherent secreted mucus layer, the glycocalyx and secretory goblet cells. Taken from R.D. Pullan, G.A. Thomas, M. Rhodes, R.G. Newcombe, G.T. Williams, A. Allen, and J. Rhodes,* Gut, *1994, 35, 353 with permission*

layers up to 700 μm thick under normal physiological conditions (Figure 1). This mucus layer is the main external barrier providing physical protection of the underlying mucosal cells. Due to the sensitivity of the secreted mucus layers to dehydration, they are easily and rapidly lost during conventional preparation methods for tissue samples[11]. Accordingly, the collection of intact mucosal samples needs to be carefully controlled, in order to obtain a physiologically relevant mucus thickness value. In spite of these difficulties, mucus thickness measurement is the best and most direct indicator of normal and pathological function and has great value in screening for mucosal disease.

The protective role of the mucus barrier has focussed attention on the range of molecular components encountered and considerable effort has been made to identify their molecular characteristics and function, especially potential interactions with each other within the mucus barrier system. Study of the secreted mucus layer and the glycocalyx composition have attracted a wide range of approaches[2,4-6,8-10,12-15]

The GI tract is characterised by a single layer of epithelial cells which synthesise and secrete mucus. The secretory, or goblet cells are readily identified by their mucus filled thecae (Figure 2)[16]. The secretory nature of the epithelium is maintained throughout the GI tract from stomach to rectum and shows a maturation profile at each location culminating in cell death and shedding at external mucosal surfaces[16]. Detailed information concerning the origin of the different populations of mucosal cell types and their stem cell origins has been reported and this correlates well with mucus production and interaction with both the innate and adaptive immune systems and the enteric microflora[1,2,9]. In the small intestine, a population of secretory cells, the Paneth cells, produce mucus and also contain intracellular granules rich in antimicrobial molecules which contribute to the maintenance of sterility in the mucus gel immediately adjacent to the mucosal surface[17].

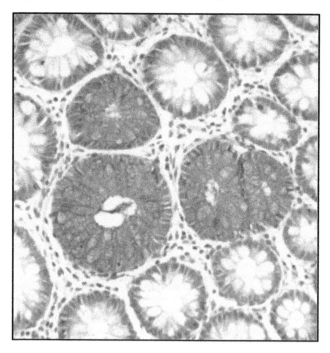

Figure 2 *Mucus stored in goblet cell thecae. Staining of the colonic mucosa with the mild periodic acid–Schiff (mPAS) reaction stains non-O-acetylated sialic acids and demonstrates the location of the mucus before secretion. Taken from A.P. Corfield,* The Biochemist, *2011, 10, with permission*

3 THE MUCIN GENE FAMILY

The family of mucins, (MUC genes) having 18 or more members is expressed on a tissue specific basis. The mature expressed forms fall into three groups, the secreted, gel-forming mucins; the secreted non-gel forming mucin MUC7 and the membrane-associated forms. The characteristics of MUC8 remain unclear, but no gastrointestinal expression has been reported (Table 1)[2,4,5,8].

The mucins are flexible linear polymers and the mucins in each group can be characterised by their peptide domain organisation. This molecular organisation is closely linked with the generation of networks for the secreted forms and the composition of the cell surface glycocalyx for membrane associated mucins. All secreted mucins show a homo-oligomeric structure with disulphide bridges linking mucin monomers at cysteine-rich or cystine-knot sequences and von Willebrand C and D domains located at N-and C-terminal peptide domains[18]. These peptide sequences flank the variable number of tandem repeat (VNTR) domains, which carry the oligosaccharide chains. MUC2, the major human intestinal mucin has N-terminus von Willebrand D1, D2, D'D3, cystine rich D, small PTS (proline, threonine, serine), cystine rich D, large PTS, C terminal von Willebrand D4, von Willebrand B, von Willebrand C and finally cysteine knot domain (CK)[19]. The translated MUC2 product is rapidly dimerised through CK disulphide bridges.

Table 1 *The Mucin Gene Family. Chromosomal location, tandem repeat size and gastrointestinal expression*

Mucin	Chromosome	Tandem repeat size (amino acids)	Gastrointestinal expression
Secreted mucins – gel forming			
MUC2	11p15.5	23	Small intestine, colon
MUC5AC	11p15.5	8	Stomach
MUC5B	11p15.5	29	Salivary glands, gallbladder
MUC6	11p15.5	169	Stomach, duodenum, gallbladder, pancreas
MUC19	12q12	19	Sublingual gland, submandibular gland
Secreted mucins – non-gel forming			
MUC7	4q13 – q21	23	Salivary glands
MUC8	12q24.3	13/41	Not detected
MUC9	1p13	15	Not detected
Membrane-associated			
MUC1	1q21	20	Stomach, gallbladder, pancreas, duodenum, colon
MUC3A/B	7q22	17	Small intestine, colon, gall bladder, duodenum
MUC4	3q29	16	Colon, stomach
MUC12	7q22	28	Colon, small intestine, stomach, pancreas
MUC13	3q21.2	27	Colon, small intestine, stomach
MUC15	11p14.3	none	Small intestine, colon, foetal liver
MUC16	19p13.2	156	Peritoneal mesothelium
MUC17	7q22	59	Small intestine, colon, duodenum, stomach
MUC20	3q29	18	Colon
MUC21	6p21	15	Colon

This occurs in the endoplasmic reticulum[20], before migration to the Golgi apparatus where mucin type O-glycosylation of the PTS domains occurs. A further molecular rearrangement takes place in the trans-Golgi network where trimers are formed[21]. These MUC2 structures are concentrated in the goblet cell vesicles. In accord with von Willebrand factor oligomerisation the events which govern packing and secretion depend on pH and Ca^{2+} ion concentration[22]. Trimerisation of MUC2 through the N-terminal peptide domains plays a vital role in the events leading to the formation of mucus networks[22]. This oligomerisation model presents a possible mode of action to explain the large increase in volume characteristic for mucins on secretion.

The membrane-associated mucins form the largest group of mucins (Table1) and are typical monomeric, membrane anchored glycoproteins, which do not form gels[18, 23-26]. They are major components of the glycocalyx in apical cell membranes on mucosal surfaces and have a peptide domain structure which reflects their functions at this location. The mucin domains extend into the extracellular space and present an O-glycan glycoarray at the cell surface enabling multiple host interactions with the external environment. Functional peptide domains include the the sea-urchin sperm protein, enterokinase and agrin (SEA) module and epidermal growth factor (EGF)-like domain.[27] The SEA module contains a peptide cleavage site, forming a non-covalent complex[28] and enabling the release of the large extracellular mucin domain which can be detected in the mucus gel layer[29-31]. As shown in Table 1 the major membrane –associated mucins in the GI tract are represented by MUC1, MUC4, MUC12, MUC13, MUC17, MUC20 and MUC21. An additional source of secreted, soluble mucins arises due to alternative splicing and generates molecules with no transmembrane or cytoplasmic peptide domains[32-34]. The main part of this data is available for MUC1 and only limited understanding of the biological role of these splice variant forms exists at present. Due to their smaller molecular size they do not co-purify with the large mucins and must be detected using specific reagents and techniques.

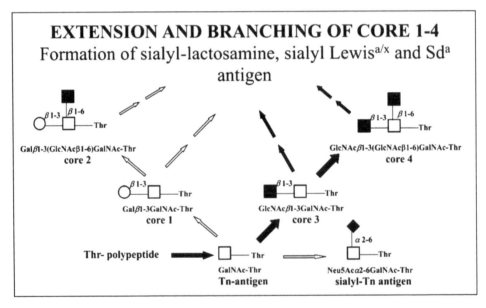

EXTENSION AND BRANCHING OF CORE 1-4
Formation of sialyl-lactosamine, sialyl Lewis[a/x] and Sd[a] antigen

Figure 3 *Glycosylation pathways in the GI tract. Biosynthesis of core mucin type O-glycans is shown. major pathways in the GI tract are shown with black arrows and minor pathways by open arrows. The arrows indicate further extension of the core 1-4 structures through other pathways to yield sialyl-lactosamine, sialyl-Lewis[a/x] and Sd[a] antigen. Glycan structures are shown by formulae and symbols. Taken from Corfield, A.P. (2011) The Biochemist 10-15, with permission*

4 MUCIN GLYCOSYLATION

The glycosylation of mucins is focussed on the VNTR, PTS domains found in the central part of the molecules. . Serine and threonine residues provide sites for glycan linkage through N-acetyl-D-Galactosamine, yielding the O-glycan, mucin-type oligosaccharides. Glycans based on Core 3 are most common in gastrointestinal mucins and these may be extended by N-acetyl-lactosamine units and terminated with fucose, sialic acid or ester sulphate[5,14,35-39]. In addition, a smaller number of N-glycans are found and these are associated with processing of the mucin peptide during biosynthesis[13,40].

A further post translational modification with implications for mucin function is C-mannosylation of the CysD domains. The covalent, C–C linkage binding of an α-manno-pyranosyl residue to the indole C2 carbon atom of the first Tryptophan in WXXW sequences,[41]. Although functional details are still scanty C-mannosylation is thought to act at the protein folding, sub-cellular localisation and trafficking levels at the ER-Golgi interface[42,43]. The Cys domains are found in the secreted mucins MUC2 (two copies), MUC5B (seven copies) and MUC5AC (nine copies) in man and numerous secreted mucins and other *O*-glycoproteins in animals[44]. Studies with recombinant MUC5B have indicated that Cys domains must be *C*-mannosylated to ensure correct maturation and secretion of the mucin. If this does not occur the recombinant molecules are retained in the ER where they induce ER stress.

Glycosylation is a tissue specific event and this is reflected in the glycan structures reported for mucins from different tissue sources and the metabolic pathways generating the O-glycans are well known (Figure 3)[2,4,35,37-39]. The human gastrointestinal tract shows a variation in the glycosylation of MUC2 expressed at each region, from the small intestine to the rectum[45,46]. Mucosal disease events also influence glycosylation. As an example the MUC5AC gene is expressed in both stomach and trachea in man, but the glycosylation for the mucin products is different in each organ[47,48]. Glycosylation patterns also impinge on the interaction of the microflora with the host in achieving efficient mucosal protection[49]. Both infection and inflammation processes lead to changes that influence susceptibility to pathogens[2-4,50-52]. A number of glycosylation linked biomarkers relate to GI disease and have been identified on the relevant metabolic pathways[2,4,35,37-39]

5   FORMATION OF MUCUS GELS AT THE GASTROINTESTINAL MUCOSAL SURFACE. A DYNAMIC SYSTEM

The formation of a mucosal surface mucus gel is a primary facet of the innate, defensive barrier. This gel barrier is required at all times in order to provide effective protection. At the same time there will be mucus turnover as the contact with the external environment will lead to disruption and degradation of the mucus barrier components, again on a continuous basis. A positive balance between creation of new and intact mucus gel and the turnover and elimination of the degraded mucus is required and is the norm for the mucosal surface *in vivo*. These events include mucin biosynthesis, dimerisation, trimerisation and oligomerisation and network formation on secretion. Physical disruption of the mucus gel and concomitant enzymatic degradation of the glycan chains and the peptide backbone represent the mucus turnover events.

Among the mucin peptide biosynthetic and degrading enzymes are those which generate and cleave disulphide bridges found within the mucin monomers themselves and also between monomers to yield dimers, trimers and oligomers[18,22,42,53-56]. The generation of mucin gel networks at the mucosal surface relies primarily on dimer and trimer formation, mediated through disulphide bridges between specific peptide domains[22, 56]. Other proteins present in mucosal secretions may also contribute to the creation or maintenance of mucus gels through molecular crosslinks. Examples include the trefoil peptides, gastrokines, transferrin, secretory IgA and others[6,39].

There is a considerable contrast in the temporal schedules for mucin synthesis and secretion. Biosynthesis of MUC5AC has been estimated at around 2 hours[57], while secretion and hydration occur in the milli-second to second span[58]. The secretory granules exhibit a low pH of 5.2, in contrast to the other sites of biosynthesis, pH 7.2 in the ER and pH 6.0 in the trans-Golgi network. In addition, a high intragranular Ca^{2+} level is found[22]. Thus, lower pH and higher Ca^{2+} levels appear to be necessary for granular packing. The dramatic increase in mucin volume which occurs on secretion has been partly explained on the basis of an ionic gradient, with an exchange of divalent Ca^{2+} for monomeric Na^+ and a lower pH of 5.2[22,56].

Several examples underline the adaptation of this process to suit the individual needs of the mucosal barrier at specific organ/tissue sites. Mucin glycoforms of the same gene product can arise from the same gland[59] or in neighbouring goblet cells[60] or, in contrast, MUC gene products may be synthesised at different tissue sites within the same organ, such as MUC5AC and MUC6 in the stomach and where discrete layers of each mucin can be detected in the secreted mucus gel[61].

6 MUCINS AND THE INTESTINAL MICROFLORA

The interaction of the host mucosal epithelium and its' mucus barrier with the indigenous microflora has long been a focus of attention. Both normal and pathological events are related to the function of these interactions and considerable efforts have been made to better understand the mechanisms involved. A two layer mucus barrier has been detected, the inner adherent gel is free of bacteria, which are only found in the outer, thicker layer under normal conditions[62]. Both are comprised of MUC2 and the outer layer is formed from the inner through the action of proteolytic and glycosidase activities from host and microflora. Thus gastrointestinal MUC2 gels in the mucosal barrier prevent the microflora from contacting and infecting the epithelium. The glycan array presented by MUC2 serves as attachment sites for the colonic microflora and suggests a basis for the potential selection of strains. Changes in the glycosylation of MUC2 may lead to mucosal disease and inflammation[1,9]. As discussed below the MUC2 O-glycans may also serve as an energy source[9]

The gut microflora is implicated in gut angiogenesis and glycan legislation[49, 63-65]. The relationship of the host with the microflora must account for the dynamic nature of the gut microbiome. The range of symbiotic, commensal and pathogenic strain present in the gut lead to a mutualism, which explains the bacterial genetic diversity, the capacity for rapid growth and high population density. These interactions are a driving force for the co-evolution of an effective host-microflora relationship[63]. There are well established patterns of bacterial colonisation associated with the gastrointestinal tract, e.g. Lactobacillus sp. in the stomach and E. coli sp. in the colon, and also normal development with discrete colonisation patterns associated with the neonate, lactation, weaning and adulthood. Similarly, in disease discrete bacterial strains are identified, such as *Helicobacter pylori* in the stomach[66] and Intestinal Bowel Disease[1]. Molecular analysis of the factors that determine the nature of these relationships must include the repertoire of glycans expressed by the host and the bacterial ability to access and utilize them. Furthermore, the need to evade pathogenic microbes and create and sustain a symbiotic relationship with indigenous bacteria is a strong selective pressure to create the capacity for glycan diversity[67,68]. These selective pressures have led to the co-evolution of diverse host glycotopes in the mucosal barrier and microbial systems to manipulate them. They include glycosidases, monosaccharide permeases and monosaccharide intermediary metabolic enzymes. Such evolution represents a capacity to generate structural diversity in glycans through non-template based mechanisms. This facilitates the ability of the host to accommodate dynamic fluxes in the composition of the gut flora and the rapid evolution of individual species. Evidence of *in vivo* host selection for bacteria that can degrade specific, individual glycotopes exists. This has been demonstrated in the faecal flora of individuals with defined blood group status and who only degrade their host blood group. Those with other blood groups show a different pattern of hydrolases, specific for each group. The transfer of microflora between different blood group individuals is incompatible[69,70]. Intestinal bacterial species require fermentable carbohydrates as an energy source and therefore require the ability to degrade glycoconjugates through the action of specific glycosidases. This gives them an ecological advantage, i.e. direct access to a host derived energy source. However, only some strains have all enzymes to do this and these are known as the mucin oligosaccharide degrader (MOD) strains. These strains generate monosaccharides and can thus recruit other organisms that will use them. These observation have led to the hypothesis known as glycan legislation (Figure 4)[49,63,71].

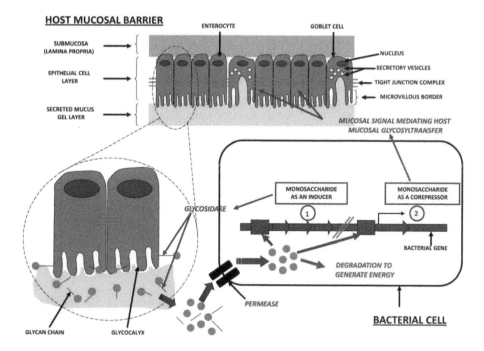

Figure 4 *Glycan legislation. The glycobiological interactions of human enteric bacteria with the host is represented. Taken from ref. 80, with permission*

The glycosylation of the human gastrointestinal barrier has recently been shown to express specific glycan structures in a differential manner throughout the intestinal tract [45, 46, 72] This implies a biological role for these glycans as part of normal mucosal protection.

Glycan legislation has not been directly tested in man, but faecal flow to the distal colon can be removed by diversion surgery and provides a way of testing this hypothesis. Screening of biopsy samples taken at operation and then four months later at final closure allowed the glycosylation patters in the mucosa to be tested in the absence of the main faecal microflora. Removal of faecal flow led to regulation of selected glycan structures in the host mucosa, thus demonstrating that glycan legislation operates in man. This process can account for short-term responses to the requirements of the microflora without the need to alter the expression of major barrier glycoprotein genes, while the differential expression of glycan structures implies a biological role in host-bacterial symbiotic interactions[14].

7 MUCUS TURNOVER, INTEGRATION WITH MUCIN DYNAMICS

The turnover of mucus gels is closely linked with the features already outlined above. Secreted mucus is continually exposed to dietary constituents, the microflora and host factors, all of which act together to degrade the barrier[5,6,73,74]. The combined action of peptidases, proteases and glycosidases generate fragments which form complexes with agents such as the bile salts[75,76]. These interactions are particularly relevant in the small intestine, where the barrier is thinnest, but where major nutrient uptake occurs. Physiological surfactants including the bile salts and phospholipids allow protein denaturation, lipolysis, product transport in mixed micelles and contribute to mucosal defence.[77, 78]

Mucus turnover may be investigated using a variety of techniques, including the action of glycosylation inhibitors[14,60,79,80]. In this case, induction of apoptosis and block of cellular proliferation is found with a depletion of the mucus layer and a change in mucin glycosylation[14, 80]. Gene array analysis revealed that mucus turnover is only one metabolic event regulated by glycosylation pathways, with transcription, signal transduction and proliferation showing the major changes[80].

In conclusion, the range of events mediating mucus turnover are integrated to accommodate the dynamic requirements of the management of these complex molecules. It is clear that further probing of mucus turnover will evolve in these areas of research.

References

1   R.J. Xavier and D.K. Podolsky, *Nature,* 2007, **448**, 427.
2   M.A. McGuckin, S.K. Linden, P. Sutton and T.H. Florin, *Nature Reviews Microbiology,* 2011, **9**, 265.
3   S.K. Linden, T.H. Florin and M.A. McGuckin, *PloS one,* 2008, **3**, e3952.
4   S.K. Linden, P. Sutton, N.G. Karlsson, V. Korolik and M.A. McGuckin, *Nature Mucosal Immunology,* 2008, **1**, 183.
5   I. van Seuningen, *The Epithelial Mucins: Structure/Function. Roles in Cancer and Inflammatory Diseases.* Research Signpost, Kerala, India, 2008.
6   A.P. Corfield, D. Carroll, N. Myerscough and C.S. Probert, *Front. Biosci.* 2001, **6**, D1321.
7   M.E. Johansson, J.M. Larsson and G.G. Hansson, G. C. *Proc. Natl. Acad. Sci. USA,* 2011, **108** *Suppl 1,* 4659.
8   A.P. Moran, A. Gupta and L. Joshi, *Gut,* 2011, **60**, 1412.

9   M.E.V. Johansson, D. Ambort, T. Pelaseyed, A. Schutte, J.K. Gustafsson, A. Ermund, D.B. Subramani, J.M. Holmen-Larsson, K.A. Thomsson, J.H. Bergstrom, S. van der Post, A.M. Rodriguez-Pineiro, H. Sjovall, M. Backstrom and G.C. Hansson, *Cell. Mol. Life Sci.,* 2011, **68,** 3635.

10   R.A. Cone, *Advanced Drug Delivery Reviews,* 2009, **61,** 75.

11   V. Strugala, A. Allen, P.W. Dettmar and J.P. Pearson, *Proc. Nutr. Soc.,* 2003, **62,** 237.

12   C. Atuma, V. Strugala, A. Allen and L. Holm, *Am. J. Physiol. Gastrointest. Liver Physiol.,* 2001, **280,** G922.

13   M.A. Hollingsworth and B.J. Swanson, *Nat. Rev. Cancer.,* 2004, **4,** 45.

14   G. Patsos and A. Corfield, *Biological Chemistry,* 2009, **390,** 581.

15   P. Gniewek and A. Kolinski, *Biophysical Journal,* 2012, **102,** 195.

16   R.D. Specian and M.G. Oliver, *Am. J. Physiol.,* 1991, **260,** C183.

17   T.S. Stappenbeck, *Gastroenterology,* 2009, **137,** 30.

18   J.K. Desseyn, D. Tetaert and V. Gouyer, *Gene,* 2008, **410,** 215.

19   J.R. Gum, J.W. Hicks, N.W. Toribara, B. Siddiki and Y.S. Kim, *J. Biol. Chem.,* 1994, **269,** 2440.

20   M.E. Lidell, M.E. Johansson, M. Morgelin, N. Asker, J.R. Gum, Jr., Y.S. Kim and G.C. Hansson, *Biochem. J.,* 2003, **372,** 335.

21   K. Godl, M.E. Johansson, M.E. Lidell, M. Morgelin, H. Karlsson, F.J. Olson, J.R. Gum, Jr., Y.S. Kim and G.C. Hansson, *J. Biol. Chem.,* 2002, **277,** 47248.

22   D. Ambort, M.E. Johansson, J.F. Gustafsson, H.E. Nilsson, A. Ermund, B.R. Johansson, P.J. Koeck, H. Hebert and G.C. Hansson, *Proc. Natl. Acad. Sci. USA,* 2012, **109,** 5645.

23   N. Jonckheere and I. Van Seuningen, *Biochimie,* 2010, **92,** 1.

24   S. Bafna, S. Kaur and S.K. Batra, *Oncogene,* 2010, **29,** 2893.

25   N. Jonckheere, and I. Van Seuningen, *Critical Reviews in Oncogenesis,* 2008, **14,** 177.

26   C.L. Hattrup and S.J. Gendler, *Ann. Rev. Physiol.,* 2008, **70,** 431.

27   D.H. Wreschner, M.A. McGuckin, S.J. Williams, A. Baruch, M. Yoeli, R. Ziv, L. Okun, J. Zaretsky, N. Smorodinsky, I. Keydar, P. Neophytou, M. Stacey, H.H. Lin and S. Gordon, *Protein Sci.,* 2002, **11,** 698.

28   B. Macao, D.G. Johansson, G.C. Hansson and T. Hard, *Nat. Struct. Mol. Biol.,* 2006, **13,** 71.

29   A. Thathiah, C.P. Blobel and D.D. Carson, *J. Biol. Chem.,* 2003, **278,** 3386.

30   A. Thathiah and D.D. Carson, *Biochem. J.,* 2004, **382,** 363.

31   S.J. Williams, D.H. Wreschner, M. Tran, H.J. Eyre, G.R. Sutherland and M.A. McGuckin, *J. Biol. Chem.,* 2001, **276,** 18327.

32   S. Zrihan-Licht, H.L. Vos, A. Baruch, O. Elroy-Stein, D. Sagiv, I. Keydar, J. Hilkens and D.H. Wreschner, *Eur. J. Biochem.,* 1994, **224,** 787.

33   S.J. Williams, M.A. McGuckin, D.C. Gotley, H.J. Eyre, G.R. Sutherland and T.M. Antalis, *Cancer Res.,* 1999, **59,** 4083.

34   S.J. Williams, S., D.J. Munster, R.J. Quin, D.C. Gotley and M.A. McGuckin, *Biochem. Biophys. Res. Commun.,* 1999, **261,** 83.

35   G. Patsos and A. Corfield, in *The Sugar Code. Fundamentals of Glycoscience,* ed. H.-J. Gabius, H.-J., Wiley-VCH Verlag GmbH & Co. KGaA, Weinheim, Germany, 2009, p. 111.

36   I. Brockhausen, *Adv. Exp. Med. Biol.,* 2003, **535,** 163.

37   I. Brockhausen, *EMBO Rep.,* 2006, **7,** 599.

38   L. Xia, *Methods in Enzymology,* 2010, **479,** 123.

39   S. Wopereis, D.J. Lefeber, E. Morava and R.A. Wevers, *Clin. Chem.,* 2006, **52,** 574.

40   G. Theodoropoulos and K.L. Carraway, *J. Cell Biochem.,* 2007, **102,** 1103.

41   J. Hofsteenge, D.R. Muller, T. de Beer, A. Loffler, W.J. Richter and J.F. Vliegenthart, *Biochemistry,* 1994, **33,** 13524.

42   D. Ambort, D., S. van der Post, M.E. Johansson, J. Mackenzie, E. Thomsson, U. Krengel and G.C. Hansson, *Biochem. J.,* 2011, **436,** 61.

43   J. Perez-Vilar, S.H. Randell and R.C. Boucher, *Glycobiology,* 2004, **14,** 325.

44   J.L. Desseyn, *Molecular Phylogenetics and Evolution,* 2009, **52,** 284.

45   C. Robbe, C. Capon, B. Coddeville and J.C. Michalski, *Biochem. J.,* 2004, **384,** 307.

46   C. Robbe, C. Capon, E. Maes, M. Rousset, A.   Zweibaum, A., J.P. Zanetta and J.C. Michalski, *J. Biol. Chem.*, 2003, **278**, 46337.

47   C. de Bolos, F.X. Real and A. Lopez-Ferrer, *Front Biosci.*, 2001, **6**, D1256.

48   D.J. Thornton, I. Carlstedt, I., M. Howard, P.L. Devine, M.R. Price and J.K. Sheehan, *Biochem. J.*, 1996, **316**, 967.

49   L.V. Hooper and J.I. Gordon, *Glycobiology*, 2001, **11**, 1R.

50   M.A. McGuckin, A.L. Every, C.D. Skene, S.K. Linden, Y.T. Chionh, A. Swierczak, J. McAuley, S. Harbour, M. Kaparakis, R. Ferrero and P. Sutton, *Gastroenterology*, 2007, **133**, 1210.

51   J.L. McAuley, S.K. Linden, C.W. Png, R.M. King, H.L. Pennington, S.J. Gendler, T.H. Florin, G.R. Hill, V. Korolik and M.A. McGuckin, *J. Clin. Invest.*, 2007, **117**, 2313.

52   A. Sturm, A and A.U. Dignass, *World J. Gastroenterol.* 2008, **14**, 348.

53   S.L. Bell, G. Xu, I.A. Khatri, R. Wang, S. Rahman and J.F. Forstner, *Biochem. J.*, 2003, **373**, 893.

54   V.M. Leitner, G.F. Walker and A. Bernkop-Schnurch, *Eur. J. Pharm. Biopharm.*, 2003, **56**, 207.

55   M. Kesimer, C. Ehre, K.A. Burns, C.W. Davis, J.K. Sheehan and R.J. Pickles, *Mucosal Immunology*, 2012 **5**, doi:10.1038/mi.2012.81 (in press).

56   M. Kesimer, A.M. Makhov, J.D. Griffith, P. Verdugo and J.K. Sheehan, *Am. J. Physiol. Lung Cell Mol. Physiol.*, 2010, **298**, L15.

57   J.K. Sheehan, S. Kirkham, M. Howard, P. Woodman, S. Kutay, C. Brazeau, J. Buckley and D.J. Thornton, *J. Biol. Chem.*, 2004, **279**, 15698.

58   P. Verdugo, *Ann. Rev. Physiol.*, 1990, **52**, 157.

59   E.C. Veerman, P.A. van den Keijbus, K. Nazmi, W. Vos, J.E. van der Wal, E. Bloemena, J.G. Bolscher and A.V. Amerongen *Glycobiology*, 2003, **13**, 363.

60   G. Huet, V. Gouyer, D. Delacour, C. Richet, J.P. Zanetta, P. Delannoy and P. Degand, *Biochimie*, 2003, **85**, 323.

61   S.B. Ho, K. Takamura, R. Anway, L.L. Shekels, N.W. Toribara and H. Ota, *Dig. Dis. Sci.*, 2004, **49**, 1598.

62   M.E. Johansson, M. Phillipson, J. Petersson, A. Velcich, L. Holm and G.C. Hansson, *Proc. Natl. Acad. Sci. USA*, 2008, **105**, 15064.

63   L.V. Hooper and J.I. Gordon, *Science*, 2001, **292**, 1115.

64   F. Backhed, R.E. Ley, J.L. Sonnenburg, D.A. Peterson and J.I. Gordon, *Science*, 2005, **307**, 1, 915.

65   J.L. Sonnenburg, L.T. Angenent and J.I. Gordon *Nature Immunology* 2004, **5**, 569.

66   Y. Rossez, E. Maes, T. Lefebvre-Darroman, P. Gosset, C. Ecobichon, C.; M. Joncquel Chevalier Curt, I.G. Boneca, J.C. Michalski and C. Robbe-Masselot, *Glycobiology*, 2012, **22**, 1193.

67   P. Gagneux and A. Varki, A. *Glycobiology*, 1999, **9**, 747.

68   A. Varki, *Glycobiology*, 1993, **3**, 97.

69   L.C. Hoskins, M. Agustines, W.B. McKee, E.T. Boulding, M. Kriaris and G. Niedermeyer, *J. Clin. Invest.*, 1985, **75**, 944.

70   L.C. Hoskins and E.T. Boulding, *J. Clin. Invest.*, 1976, **57**, 63.

71   L.V. Hooper, M.H. Wong, A. Thelin, L. Hansson, P.G. Falk, and J.I. Gordon, *Science*, 2001, **291**, 881.

72   C. Robbe-Masselot, E. Maes, M. Rousset, J.-C. Michalski and C. Capon, *Glycoconjugate Journal*, 2009, **26**, 397.

73   M. Andrianifahanana,N. Moniaux and S.K. Batra, *Biochim. Biophys. Acta*, 2006, **1765**, 189.

74   B. Deplancke and H.R. Gaskins, *Am. J. Clin. Nutr.*, 2001, **73**, 1131S.

75   A. Macierzanka, N.M. Rigby, A.P. Corfield, N. Wellner, F. Bottger, E.N.C. Mills and A.R. Mackie, *Soft Matter*, 2011, **7**, 8077.

76   T.S. Wiedmann, H. Herrington, C. Deye and D. Kallick, *Chem. Phys. Lipids*, 2001, **112**, 81.

77   J.M. Barrios and L.M. Lichtenberger, *Gastroenterology*, 2000, **118**, 1179.

78   J. Gass, H. Vora, A.F. Hofmann, G.M. Gray and C. Khosla, *Gastroenterology*, 2007, **133**, 16.

79 W. Morelle, L. Stechly, S. Andre, I. van Seuningen, N. Porchet, H.J. Gabius, J.C. Michalski and G. Huet, *Biol. Chem.,* 2009, **390,** 529.

80 G. Patsos, V. Hebbe-Viton, C. Robbe-Masselot, D. Masselot, R. San Martin, R. Greenwood, C. Paraskeva, A. Klein, M. Graessmann, J.C. Michalski, T. Gallagher and A. Corfield, *Glycobiology,* 2009, **19,** 382.

7

VISCOMETRY, ANALYTICAL ULTRACENTRIFUGATION AND LIGHT SCATTERING PROBES FOR CARBOHYDRATE STABILITY

S.E. Harding

National Centre for Macromolecular Hydrodynamics, School of Biosciences, University of Nottingham, Sutton Bonington, LE12 5RD, UK
steve.harding@nottingham.ac.uk

1 INTRODUCTION

Hydrodynamic probes vary in their degree of sophistication and capability, although between them they provide a complementary set of tools for the assessment of polymer stability. This is particularly relevant for large carbohydrate polymer assemblies, whether it be resistance to chain scission or resistance against aggregative phenomena. All these probes involve motions of the polymer in solution – either in terms of the effect on the bulk flow properties in response to a shearing force (viscosity), or the motions of the polymer through a solution in response to a force field. The latter can be thermal/stochastic (light scattering) or gravitational (chromatographic flow and ultracentrifugation).

In this chapter we highlight some of the information obtainable from the modern implementation of these well-established methods, some limitations and the virtue of using these methods in combination. The types of information that can be reasonably expected are (i) heterogeneity information; (ii) molecular weight (molar mass) averages and distributions; (iii) the extent and reversibility of aggregation and degradative phenomena; (iv) conformation and conformational flexibility. In the following chapter by Morris *et al.* the application of some of these methods to the study of the stability of one important class of carbohydrate polymer – the pectins - is considered. We cover only an overview of the techniques here although some key references giving the necessary detail behind them are indicated.

2 VISCOMETRY

This is the simplest – and least expensive - of the hydrodynamic probes and can provide basic information about conformation, conformation change (e.g. in response to thermal treatment), and, for non-spheroidal particles molecular weight and the state of aggregation/degradation. Some key viscosity parameters (see ref. 1 and references therein) are the absolute viscosity (or ratio of the shear stress to shear rate), η, the relative viscosity (ratio of the solution viscosity at a concentration c (g/ml), to the solvent viscosity) $\eta_r = \eta_{solution}/\eta_{solvent}$ and the reduced viscosity η_{red} (ml/g) $= (1 - \eta_r)/c$. To eliminate the effects of non-ideality the reduced viscosity – or alternatively the inherent viscosity $= \ln\eta_{rel}/c$ – is extrapolated to zero concentration (via the Huggins and Kramer plots respectively) to yield the intrinsic viscosity $[\eta]$ which is an intrinsic function of the properties of the macromolecule. If the viscometer is sensitive enough then a concentration extrapolation may not be necessary and $[\eta] \approx \eta_{red}$ at high dilution. A better approximation is given by the Solomon-Ciuta expression[2]

$$[\eta] \simeq (1/c). [2\{\eta_{sp} - \ln(\eta_{rel})\}]^{1/2} \tag{1}$$

The relative viscosity itself can be measured using conventional capillary or "Ostwald" viscometers – controlled to a temperature of at least $\pm 0.01^{\circ}$C – provided there is no shear dependence or "non-Newtonian" behaviour of the measured $[\eta]$. If there is, then a rolling ball viscometer of the type for example manufactured by Anton-Paar Instruments (Graz, Austria) using a steel ball (coated with silane, teflon or other inert material) rolling through solution in a capillary at different set roll angles - with extrapolation to zero angle (where the shear rate→0), may be useful[3]. Alternatively conventional "cone and plate" rheometers – where shear stresses and shear rates are measured directly - can be used to correct for any shear effects[4]. Another potential complication is one of coil overlap above a certain concentration, usually represented as c*. An approximate relation linking c* and $[\eta]$ has been given[5]:

$$c^* \approx 3.3/[\eta] \tag{2}$$

The intrinsic viscosity depends on the conformation, flexibility and volume of the polymer. The strong hydrodynamic interactions between the segments of a polymer mean that effectively a polymer chain can normally be considered as a single hydrodynamic particle as considered for example by Tanford and others[6], with solvent between the segments of a polymer moving with the polymer. Departure from expected behaviour is sometimes explained in terms of partial draining of the solvent[7].

The dependence of the intrinsic viscosity on molar mass is strongly dependent on the conformation and flexibility, and the simplest representation of this is the Mark-Houwink-Kuhn-Sakurada (MHKS) equation (see for example refs 8 & 9 & references therein):

$$[\eta] = K_{visc}.M^a \tag{3}$$

where a is the MHKS or "power law" coefficient, and K_{visc} is a coefficient which also depends on the conformation: both can be estimated from a plot of $\log[\eta]$ vs $\log M$. The most useful parameter is a: for a sphere $a=0$, for a (non-draining) random coil $a\sim0.5$-0.8 and for a stiff molecule $a>1.0$ with the limit $a\sim1.8$. It can be seen from these values that for spheroidal/globular molecules such as many proteins, although $[\eta]$ is very sensitive to change in conformation – as in the classical demonstration of the thermal denaturation of ribonuclease[10] – $[\eta]$ is not a useful measure of change in molar mass through degradation or aggregation. By contrast for most polysaccharides – whose conformations are generally between those of coils and rods – $[\eta]$ is a useful measure of change in M. An example of a straightforward stability evaluation using an Ostwald capillary viscometer (based on η_{red} measurements at a concentration of 1 mg/ml) for the polycationic chitosan (poly N-acetyl glucosamine – chitin – deacetylated according to specification) is shown in Figure 1a. Here the effects of different storage temperatures on the decay profiles for this material – being considered as a nasal mucoadhesive formulation[11] - were explored[12], showing that storage at 4°C rather than room temperature is desirable. Figure 1b shows an intrinsic viscosity determination using the Solomon-Ciuta equation for 3 concentrations of a lignin dissolved in dimethyl sulphoxide (DMSO)[13].

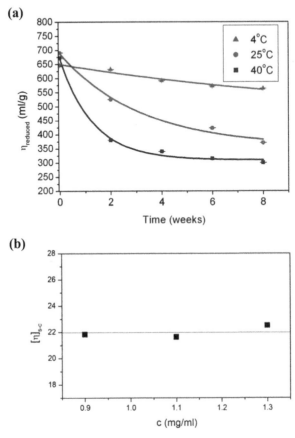

Figure 1 *(a) Intrinsic viscosity [η] or reduced specific viscosity η_{red} measurements can be used as a simple to use assay for stability of polysaccharide dispersions. Stability of chitosan (CL210) solutions (1 mg/ml in a pH 4.0, I=0.2M aqueous solvent) in response to long term storage at different temperatures[12]. Decay constants $k = 0.087$ week[-1] (4°C), 0.317 week[-1] (20°C) and 0.775 week[-1] (40°C). (b) Intrinsic viscosity measurements using the Solomon Ciuta equation for a lignin solublised in dimethyl sulphoxide[13]*

2.1 Viscometry pivotal to one of the greatest discoveries of the 20th Century

The most famous experiment using capillary viscometry on a carbohydrate polymer was that conducted by J.M. Creeth in 1947. Creeth – a PhD student at University College Nottingham supervised by D.O. Jordan and J.M. Gulland – examined the effects of extremes of pH on the relative viscosity of highly purified calf-thymus DNA, whose purity had been checked by the then fledging technique of analytical ultracentrifugation. At the extremes there was a huge drop in relative viscosity corresponding to titration/ disruption of the hydrogen bonds between the bases and break-up of the DNA molecule. This experiment – proving conclusively the existence of H-bonds in DNA – was crucial to Watson and Crick's subsequent discovery of the double helix several years later. Creeth himself produced his own early model for DNA (Figure 2b) – two chains, sugar residues and phosphate links on the outside of the molecule – and paired bases on the inside, all the essential features apart from one important detail - the helix. Although Figure 2 appeared in his Doctoral thesis of 1947[14], only Figure 2a appeared in the publication in the *Journal of the Chemical Society*[14], his supervisors regarded Figure 2b as "too speculative". This work has generally remained un-recognised until very recently[16-18].

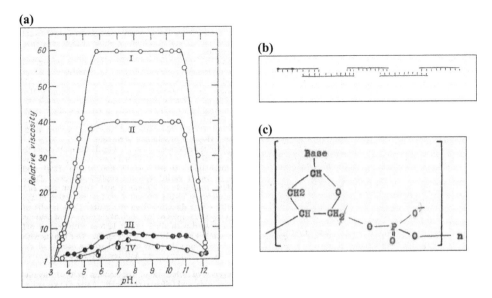

Figure 2 *(a) Plot of relative viscosity versus pH of various preparations of calf thymus DNA showing the titration out of hydrogen (H) - bonds at low and high pH*[15]*. A similar diagram appears in Creeth's PhD thesis*[14] *(b) Sketch of a model for DNA from Creeth's PhD thesis of 1947 showing two broken chains held together by H-bonds, and (c) an expanded sketch of the sugar phosphate backbone with the phosphates on the outside of the chains*

2.2 Viscosity studies on stability against irradiation and thermal degradation

An earlier study on high molar mass carbohydrate stability was on the effects of irradiation of starches. Irradiation by γ-radiation is one of the methods used to preserve foods at low doses (<10 kGy) although as with other methods of preservation (particularly thermal processing) damage to the foodstuff itself is an inevitable risk. Greenwood and MacKenzie[19] measured the effects of γ-irradiation dose (up to 100kGy) on the intrinsic viscosity of potato amylose, an α(1-4) linked glucan. A dramatic decrease was observed with increase in dose, which these researchers interpreted in terms of decrease in degree of polymerization. To make this interpretation they first of all assumed values for K_{visc} and a in the MHKS equation (Equation 1) to convert intrinsic viscosities to molar masses, and then the degree of polymerisation p is simply $M/162$, where 162g/mol is the molar mass of a glucose residue (Table 1).

Table 1 Viscometry study on the effect of γ-irradiation dose on amylose

Dose (kGy)	Intrinsic viscosity [η] ml/g	Degree of polymerization p
0	230	1700
0.5	220	1650
1	150	1100
2	110	800
5	95	700
10	80	600
20	50	350
50	40	300
100	35	250

Adapted from Greenwood and MacKenzie[19]

Bradley and Mitchell[20] also used the MHKS equation (Equation 3) to convert intrinsic viscosities to molecular weights in their comparative investigation of the kinetics of degradation of three polysaccharides, namely alginate, carboxymethylcellulose and κ-carrageenan – exposed to temperatures over 100°C. A specially designed high-temperature slit viscometer was used and zero shear viscosities were converted to intrinsic viscosities [η] using an approximate relation involving the coil overlap parameter c* (Equation 2). Then using the MHKS relation (Equation 3) and published values for K_{visc} and the power law coefficient "a" they were able to obtain approximate estimates of molar masses and from the slope of the reciprocal molar mass 1/M versus time they were also able to obtain first order rate constants for the degradation of the polymers. Figure 3 shows the degradation of a 15 mg/ml solution of κ-carrageenan at high temperature (118°C) in terms of the molar mass calculated in this way. From the kinetic curves and Arrhenius type of analysis these researchers were able to estimate the activation energies of depolymerisation for the 3 polysaccharides showing that alginate was less stable than carboxymethylcellulose or κ-carrageenan.

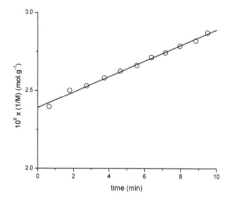

Figure 3 *Thermal degradation of κ-carrageenan at 118°C, using a specially adapted slit viscometer. Apparent molecular weights M of 15 mg/ml solutions were estimated using approximate relations from zero shear viscosity measurements. Adapted from Bradley and Mitchell[20]*

Another very useful viscometer is the Differential Pressure viscometer (see ref 1 & references therein), which uses a sensitive pressure transducer to measure the difference in pressure between solution and solvent flow through metal capillaries to evaluate the relative viscosity η_r. The flow is driven by a pump at constant rate. The three advantages of this system for carbohydrate polymers are[1] (i) it can be coupled to a size-exclusion chromatography column so that polydisperse systems can in principle be separated (ii) a concentration detector can be coupled on-line so that relative viscosities can be converted to reduced viscosities, η_{red} (ml/g), and (iii) because of the great sensitivity of the pressure transducers, low concentrations are usually only needed, often sufficient so that the approximation $[\eta] \approx \eta_{red}$ is valid. If it is not valid then the Solomon-Ciuta expression, Equation 1, can be utilised. The very low concentrations means also that the Newtonian approximation may also be more reasonable. A further advantage is that a multi – angle light scattering detector can also be coupled on-line to allow simultaneous measurement of $[\eta]$ and molar mass M, which can be very useful for carbohydrates as we shall see below. A disadvantage compared to the traditional glass capillary viscometers is that for aqueous systems in salt the metal surfaces can be more sensitive to corrosion over a period of time – appropriate coating is therefore advisable.

3 DYNAMIC LIGHT SCATTERING

Dynamic light scattering has become an increasingly popular tool for estimating the hydrodynamic radii of polymers and polymeric assemblies[21]: the instrumentation commercially available is relatively inexpensive (particularly for the single or dual angle instruments) and the method relatively rapid: this makes it particularly attractive for the investigation of changes in the molecular integrity of a system (through aggregation of degradation) with time.

From analysis of the fluctuations in intensity of scattered light caused by the (Brownian) motions of particles it is possible to estimate the translational diffusion coefficient D_t. For carbohydrate polymers and other polydisperse systems it is possible to specify both the *average* D_t – which will be a z-average[22,23], but also from the routine CONTIN[24] and related types of analysis, a *distribution* of diffusion coefficients. D_t values are popularly directly translated into hydrodynamic radii r_h using the Stokes-Einstein relation

$$r_h = k_B T / \{6\pi\eta_o D_t\} \tag{4}$$

with k_B Boltzmann's constant, T the absolute temperature and η_o the solvent viscosity. Great care has to be taken with ensuring the samples (and scattering vessels) are free of dust – choice of the appropriate filter (conventionally $\sim 0.2 - 0.5$um pore size) is important, although the modern versions of CONTIN are usually quite good at separating the signal of large particulates from the polymers in solution. With polysaccharides and other carbohydrate polymers though a word of caution is necessary particularly for the use of the single or dual angle photometers. The decay of the intensity autocorrelation function used to calculate the translational diffusion coefficient depends, for non-spherical particles on other factors most notably internal flexibilities and rotational diffusion contributions. Classically the way of eliminating these other contributions – as outlined clearly by Burchard[25] – is to extrapolate the first cumulant in the autocorrelation decay function (or equivalently the measured "apparent" translational diffusion coefficient, $D_{t,\theta}$)

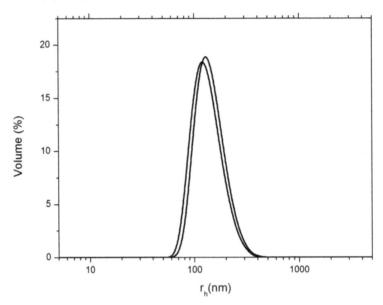

Figure 4 *Distribution of hydrodynamic radius r_h for an xanthan preparation in a phosphate chloride buffer (pH=6.8, I=0.3M) from a dynamic light scattering photometer (Malvern nano-S) at a scattering angle of 12° and a loading concentration of 0.4 mg/ml. Two measurements are shown with z-average r_h values of 120nm and 130nm respectively. At the low angle and concentration used, non-translational diffusion and non-ideality effects are assumed respectively to be negligible. From ref. 26.*

to zero angle. This is possible with conventional multi-angle instruments, but not possible with the modern fixed angle instruments. If a fixed angle instrument is being used it is essential that the scattering angle of the detector is low enough: an example of a determination on a preparation xanthan at a scattering angle $\theta = 12°$ is shown in Figure 4[26].

Care also has to be taken of concentration dependence (non-ideality) effects – although the effects are usually less severe compared to the case for intrinsic viscosity: the measured apparent translational diffusion coefficient at concentration c, $D_{t,c}$ is related to the true or "ideal" D_t ($=D^o_t$) by the relation[27,28]

$$D_{t,c} = D^o_t (1 + k_D c) \qquad (5)$$

where k_D is the concentration dependence coefficient. Usually the approximation $D_{t,c} \approx D^o_t$ is made. In cases where both the angular approximation and also the concentration dependence approximation are inappropriate, Burchard[25] introduced the concept of a dual extrapolation plot known as a *Dynamic Zimm Plot*.

It is possible also to transform distributions of D_t or r_h into distributions of molar mass using a power law relation analogous to Equation (3)[8]:

$$D_t = K_{diff}.M^{-\varepsilon} \qquad (6)$$

This is commonly employed for globular proteins[29] where K_{diff} and ε are well defined, but not so useful for carbohydrate polymers where conformations are less well defined. Classical or "static" light scattering procedures where the time averaged intensity is measured as a function of angle are by contrast better suited for providing molar mass averages and distributions and without the need for calibration standards.

4 SEC-MALS: SIZE EXCLUSION CHROMATOGRAPHY COUPLED TO MULTI-ANGLE LIGHT SCATTERING

The idea proposed over two decades ago by P.J. Wyatt[30] of coupling a classical multi-angle laser light scattering or "MALLS" photometer to size exclusion chromatography (SEC) led to one of the most significant inventions for characterizing the molar mass and molar mass distributions of polymers (n.b. the acronym MALLS has now been changed to MALS since lasers are almost universally used as the light source for light scattering experiments). SEC provides separation of molecules for a polydisperse polymer system and the fractions elute on-line into a flow cell in the MALS designed so that an instantaneous snapshot of the angular scattered intensity envelope is measured for each elution volume (V) passing through. If the concentration c(V) is also known as a function of elution volume (by coupling a concentration detector – usually refractive index based – also on-line) then the (weight average) molar mass $M_w(V)$ as a function of V can be specified and the corresponding overall weight average M_w (and also the number average M_n and z-average M_z molar masses) for the distribution - or between user specified limits of the distribution - specified. For carbohydrate polymers since the very first measurements on polysaccharides by Horton et al[31] and Rollings[32], and on mucins by Jumel et al[33] the method has become almost routine for these substances with large numbers of publications every year reporting its successful use. An early example of application to the study of glycoconjugate stability was the effect of treating mucin glycoproteins to degrading agents[34].

As with viscosity and DLS non-ideality is a consideration – measured molecular weights at a finite concentration, c, will be apparent ones, and normally an extrapolation to c=0 is necessary. However the high sensitivity of the photodiode detectors is such that single low concentrations can be injected, which are then diluted by the columns, rendering non-ideality effects small.

4.1 Effect of irradiation on guar

Jumel and colleagues[35] also used intrinsic viscosity measurements reinforced by SEC-MALS measurements of the weight average molar mass M_w and molar mass distribution, reinforced by sedimentation coefficient measurements using the analytical ultracentrifuge to evaluate the effects of gamma irradiation on the macromolecular integrity of guar, a galactomannan with a β(1-4) linked mannan backbone with α(1-4) linked galactose side chains. M_w values for the guar in response to level of irradiation (up to 9kGy) showed a 5 fold drop (Table 2) and the distributions of molar mass became correspondingly broader (Figure 5). On the basis of the molecular weight data these researchers defined an index for quantifying the incremental degree of disruption called the scission index $G_{scission}$:

$$G_{scission} = \{S_{1000} \times 100\} / \{dose(eV/g) \times g(1000 \text{ bonds})^{-1}\} \qquad (7)$$

Table 2 *Effect of γ-irradiation dose on guar (adapted from ref. 35)*

Dose (kGy)	10^{-6} x M_w g.mol^{-1}	Intrinsic viscosity [η] ml.g^{-1}	Scission index $G_{scission}$
0	2.70	1576	-
0.11	2.03	1467	10.34
0.20	2.32	1360	2.96
0.37	1.78	1360	5.00
0.50	1.87	957	3.14
0.65	1.66	1092	3.37
0.86	1.49	894	3.40
1.70	1.24	964	2.48
5.07	0.866	736	1.49
9.07	0.565	471	1.48

where 1 Gray (Gy) = 6.24 x 10^{15} eV.g^{-1}. S_{1000} is the number of scissions per 1000 glycosidic bonds, defined by $S_{1000} \sim 1000\{p_o^{-1} - p^{-1}\}$ where p_o is the degree of polymerization of non-irradiated guar. Table 2 shows the incremental G values as the dose is increased. It can be seen that as the chains become shorter and shorter, the chains become more difficult to break.

A double logarithmic Mark-Houwink-Kuhn-Sakurada plot of [η] with M_w yielded an MHKS coefficient *a* ~0.73, consistent with other estimates for guar indicating a very flexible conformation[36-38]. The conclusion was that for guar, although irradiation produces significant chain scission, it seems to have little effect on chain conformation, at least for doses below 10kGy.

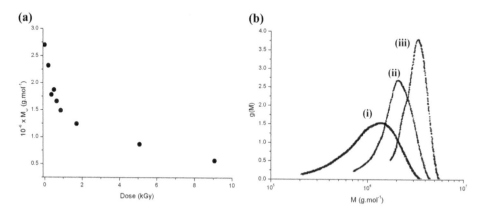

Figure 5 *SEC-MALS evaluation of the effect of γ-irradiation degradation of guar. (a) Weight average molar masses fall dramatically, and (b) the distributions also become correspondingly broader (i) 1.70kGy irradiated guar (ii) 0.20kGy irradiated (iii) non-irradiated guar. Adapted from Jumel et al[35]*

4.2 Coupling an on-line differential pressure viscometer

It is possible to couple on-line a differential pressure viscometer as described in section 2 above. This permits simultaneous determination of $[\eta](V)$ and $M_w(V)$ to be made allowing (i) a distribution of $[\eta]$ to be specified and hence weight and other average values for $[\eta]$ to be estimated (ii) a plot of $[\eta](V)$ versus $M_w(V)$ can be used to estimate the MHKS "*a*" parameter and an example is shown in Figure 6.

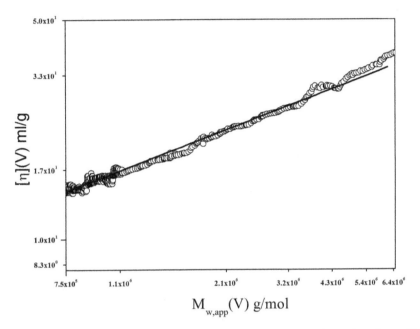

Figure 6 *Plot of $[\eta](V)$ versus $M_w(V)$ for Kikar gum polysaccharide in phosphate chloride buffer, pH=6.8, I=0.1M. The slope of this plot = (0.431±0.002). From ref. 39*

4.3 Radius of gyration

For polymers of size larger than $1/20^{th}$ of the wavelength λ of the laser light used (this is normally between 500-600nm, so particles with maximum dimension larger than $\sim$ 25nm) it is possible to also obtain from the MALS – either in batch mode[25] or coupled with SEC[30] - a useful conformation parameter from the angular dependence of the scattered light – the radius of gyration R_g - both as an average over the distribution and, with SEC-MALS, also as a function of elution volume, $R_g(V)$ vs V. This too can be related to molar mass via a power-law relation[8,9] analogous to equations (3) and (6)

$$R_g = K_{LS}.M^c \tag{8}$$

for a sphere $c\sim0.33$, for a random coil $c\sim0.5-0.6$ and for a stiff rod $c\sim1$.

4.4 Larger molar masses

A limitation of the SEC-MALS approach is the separation range of the SEC columns which tend to have an upper limit of molar mass of ~ 3 x 10^6 g/mol. For carbohydrate assemblies above this size alternative separation media can be employed, and the use of Field Flow Fractionation with appropriate membranes has been a popular choice, successfully used to characterize solubilized starches up to a molar mass[40] of ~ 5x10^8 g/mol. For some types of carbohydrate polymer – particularly cationic ones like amino-celluloses – it can be difficult finding columns (or membranes) that are sufficiently inert, and column/membrane based methods are difficult to apply to the study of reversibly associating or dissociating systems because of the variability in concentration. For these systems analytical ultracentrifugation provides a complementary approach, giving separation without the need for a separation medium.

5 ANALYTICAL ULTRACENTRIFUGATION

Like viscosity and light scattering this method is not new – T. Svedberg received the Nobel Prize for its invention in 1926 – although its modern implementation is proving a very useful tool for probing molecular integrity. At lower rotor speeds, the sedimentation force due to the centrifugal field and the backforce due to diffusion are comparable. After a considerable period of time (usually >24h) the two forces come to equilibrium: a *sedimentation equilibrium* experiment can provide absolute molar mass - primarily weight (mass) and z-averages - and molar mass distribution, analogous to what MALS can provide. The method is column (or membrane) free, but takes much longer than a MALS experiment. At much higher rotor speeds (up to 50,000 rev/min) the centrifugal force dominates, and a *sedimentation velocity* experiment can provide us with information on the physical homogeneity of a, sample, conformation and flexibility information – and if the conformation type of the polymer is known (sphere/rod/coil) an estimate of the molar mass distribution. It can also provide us with interaction information if, for example we assay for what is called "co-sedimentation" phenomena, namely different species sedimenting at the same rate[41,42].

5.1 Sedimentation equilibrium

Rotor speeds of between 2000 and 20000 rev/min are usually chosen depending on the size range of the polymer/polymer assembly, and equilibrium concentration solute distributions c(r) as a function of radial displacement r from the axis of rotation are, for carbohydrate polymer systems usually recorded using Rayleigh interference optics (Figure 7a). For mucins and other glycoproteins with sufficient protein/ uv absorbing amino acids like tyrosine, tryptophan and phenylalanine, or lignins which absorb in the uv/visible, absorbtion optics can also be used. Concentration c(r) vs radial position, r, distributions can be transformed into whole distribution apparent weight average molar masses $M_{w,app}$ using a transformation parameter known as the M* function[44] (Figure 7b), popularly incorporated into a routine known as MSTAR[43]. In addition weight average molar masses $M_{w,app}(r)$ and z-average molar masses $M_{z,app}(r)$ plots can be calculated for individual radial positions r and hence $M_{w,app}(r)$ and $M_{z,app}(r)$ as a function of concentration c(r) can be obtained, particularly important for the evaluation of reversibly aggregating systems. Such representations have recently been used to help demonstrate a reversible association in aminopolysaccharides (Figure 7c)[45].

(a)

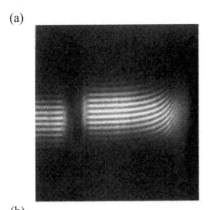

(b)

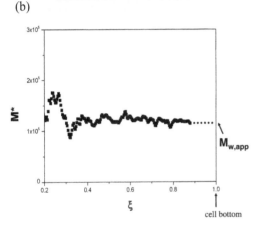

(c)

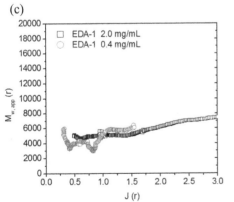

Figure 7 *Sedimentation equilibrium of carbohydrate polymers.*

(a) Rayleigh interference fringes for a bronchial mucin ($M_w \sim 6.0 \times 10^6$ g/mol), at a loading concentration of ~0.2 mg/ml and rotor speed 1967 rev/min. Each profile represents a plot of solute concentration relative to the meniscus[43].

(b) MSTAR analysis for evaluation of $M_{w,app}$ for a chitosan (loading concentration 1.0mg/ml in 0.2M acetate buffer). Fringe displacements in the vertical direction can be converted to an operational molar mass parameter $M^*(\xi)$[44] where ξ, is a normalised radial displacement (r) squared parameter: $\xi = (r^2 - r_a^2)/(r_b^2 - r_a^2)$ and where r_a and r_b the radial positions at the solution meniscus and cell base respectively. $M_{w,app}$ – the whole distribution (apparent) weight average - is obtained from the identity $M_{w,app} = M^* (\xi \rightarrow 1) = 110,000$ g/mol.

(c) It is possible also to measure the weight average molecular weights at different radial positions and hence different concentrations in the centrifuge cell: plot of "point" average molecular weight versus concentration (in fringe displacement units J(r) for an amino-cellulose at 2 different loading concentrations. The overlap of the two data-sets is symptomatic of a reversible self-association[45]

As with viscometry and light scattering, corrections for non-ideality are required, and so to evaluate M_w, M_z values from $M_{w,app}$, $M_{z,app}$, an extrapolation to c=0 is performed. At dilute solution for example:

$$(1/M_{w,app}) = (1/M_w) (1 + 2BM_wc) \qquad (9)$$

where B is the osmotic pressure 2nd virial coefficient. At very high dilution (~0.2mg/ml) the approximation $M_w \approx M_{w,app}$ can often be made (and $M_z \approx M_{z,app}$). Conversely at higher concentrations additional virial terms may be required as shown for κ-carrageenan (Figure 7c)[46].

5.2 Sedimentation velocity

The change in the radial concentration distribution profile (as with sedimentation equilibrium usually recorded using Rayleigh interference optics) with time under the influence of a strong centrifugal field (with rotor speeds usually much higher than for sedimentation equilibrium - up to 50,000 rev/min), can be transformed into a distribution of sedimentation coefficient s (in Svedberg units), or g(s) versus s plot, giving a direct measure of the heterogeneity of a sample[47]. An example is given in Figure 8 for starch[48] - the method has been used for example by Tester *et al.*[49] for the analysis of damage by ball milling processing to way, pea and maize starches. The width of a g(s) vs s profile will be affected by diffusion broadening. This can be corrected for (at least partially) to give what is known as a c(s) vs s profile, and this procedure has recently been used (reinforced by g(s) vs s measurements) to show protein-like oligomerisation of aminocelluloses[50]. Distribution software g(s) vs s or c(s) vs s is popularly incorporated into the routine SEDFIT by P. Schuck[51].

The sedimentation coefficient itself (measured at a concentration c) depends on the molar mass, shape and volume (including the effects of hydration) of the macromolecule. Like the other hydrodynamic paramers referred to above, it also needs correcting for non-ideality effects using for example this dilute solution equation to give the "ideal value"[42] s^o.

$$(1/s_c) = (1/s^o)(1+k_sc) \qquad (10)$$

A combination of s^o with M can yield the frictional coefficient ratio (ratio of the frictional coefficient of a macromolecule to a spherical particle of the same anhydrous mass) – measure of the conformation and hydration of the macromolecule. The ratio of $k_s/[\eta]$ – the "Wales van Holde ratio" is also a measure of conformation – these aspects are considered further in the following chapter in this volume by Morris *et al.* The sedimentation coefficient s^o and translational diffusion coefficient D^o_t can also be combined together (eliminating the frictional contribution) to yield an absolute measure of molar mass via the Svedberg equation[52]. For a polydisperse system Pusey has shown[53,54] that a combination of the (z-average) diffusion coefficient and (weight average) sedimentation coefficient yields the weight average molar mass M_w.

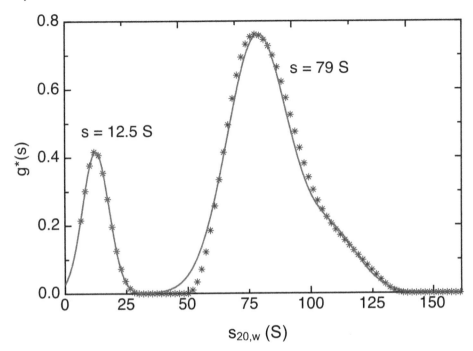

Figure 8 *Sedimentation coefficient distribution: a g(s) versus s plot for wheat starch showing two components, namely amylose (left peak) and the faster moving amylopectin, (right peak). The total sample concentration 8 mg/ml dissolved in 90% dimethyl sulphoxide. Rotor speed was 35000 rpm at a temperature of 20 °C[48].*

5.3 Combination with SEC-MALS and viscometry

An example of how analytical ultracentrifugation, SEC-MALS and viscometry can be combined together to give useful information on polysaccharide stability was given in a recent study by Patel *et al.*[52] on the effects of γ-irradiation in the molecular integrity of xyloglucans. β(1-4) linked glucan with α(1-6) linked xylose single residue side chains sometimes capped with galactomannan. Molar mass, intrinsic viscosity and sedimentation coefficient all showed significant reductions with increase in dosage (Table 3). Unlike with guar (Table 2) there seemed to be no noticeable increase in the polydispersity ratio (M_z/M_w).

As with guar, it is possible to use the data-sets for intrinsic viscosity and molecular weight together to assess the conformation using the MHKS type of relation, and in so doing Patel *et al.*[55] obtained a value of 0.55±0.03, within the expected range of a random coil (0.5-0.8). Furthermore, the sedimentation coefficient – molecular weight dataset allowed a further power-law type of analysis

$$s = K"M^b. \tag{11}$$

Table 3 *Effect of γ-irradiation dose on the physical properties of xyloglucans*

Dose (kGy)	10^{-3} x M_w g.mol^{-1}	Polydispersity M_w/M_n	Intrinsic viscosity [η] ml.g^{-1}	Sedimentation coefficient $s^o_{20,w}$ (S)^a	Persistence length L_p (nm)
0	700±5	1.1±0.1	405±35	7.21±0.03	6±1
10	270±10	1.3±0.1	210±10	4.66±0.03	6±1
20	158±3	1.4±0.1	170±10	3.10±0.04	9±1
30	127±10	1.3±0.1	140±10	3.30±0.01	6±1
40	97±10	1.3±0.1	135±5	2.82±0.04	8±1
50	60±4	1.3±0.1	100±5	2.80±0.08	6±1
70	45±3	1.1±0.1	75±5	2.61±0.02	6±2

Adapted from Patel *et al.*[55] ^{a}The superscript "o" means extrapolated to zero concentration to eliminate the effects of non-ideality. The subscript 20,w means the sedimentation coefficient has been corrected or normalized to standard conditions, namely the density and viscosity of water at 20°C. S is the Svedberg unit = 10^{-13} sec

yielding an estimate for *b* of 0.42±0.01, again between the expected limits for a random coil (0.4 – 0.5). All three datasets for M_w, [η] and $s^o_{20,w}$ could be combined into a recently developed global or HYDFIT plot procedure[56] – from combinations of the Bushin-Bohdanecky relations linking [η] with M_w and the Yamakawa-Fujii relations linking $s^o_{20,w}$ with M_w - and global minimization of a target function to yield estimates of a flexibility parameter known as the chain persistence length L_p (the practical limits are ~2nm for the highly flexible pullulan and ~ 200nm for a stiff rod like conformation like the triple helical polysaccharide schizophyllan): values obtained for this procedure for xyloglucan samples as a function of dose damage are given in (Table 3) and it appears again, as with guar (Table 2), despite chain scission there is little change in the conformational flexibility of the chain, with individual values deviating little across the whole range of molecular weight. A further example for pectins is described in the following chapter by Morris *et al.*

5.5 Molecular weight distribution from sedimentation velocity.

In the original implementation of the SEDFIT algorithm, P. Schuck[51] provided the possibility of transforming the sedimentation coefficient distribution into a molar mass distribution for paucidisperse systems such as mixtures of proteins. This has been recently extended by us to the case of quasi-continuous polydisperse systems – the hallmark of carbohydrate polymers and mucins and glycoconjugates. This method, known as the *"extended Fujita approach"* is based on an earlier method given by Fujita[57] for random coils (applied to mucins[58]) but extended to cover general conformation types[59]. An example is given for a glycoconjugate vaccine characterization in Figure 10, a system too large for characterization by SEC-MALS. Further recent examples of its use have included mucins[60].

The new "Extended Fujita" approach has now been incorporated into the SEDFIT programme. A further version of the SEDFIT algorithm will have the MSTAR procedure (section 5.1) also incorporated into it[61].

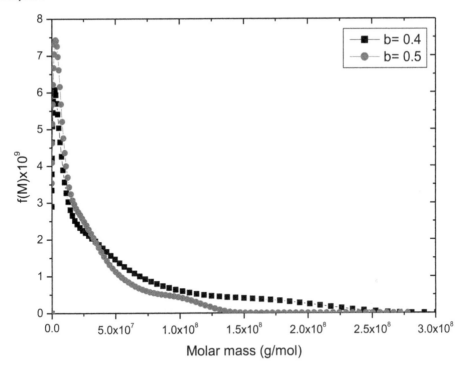

Figure 9 *Molecular weight distribution for a large glycoconjugate vaccine construct of a protein and bacterial polysaccharide obtained from sedimentation velocity data and the method described in Harding et al.[59]. Loading concentration c ~ 0.3 mg/ml. The method requires an approximate idea of the overall conformation: distributions for two reasonable selections of the power law coefficient b (Equation 11) are shown*

6 PERSPECTIVES

This short overview has reviewed the principal hydrodynamic based methods for characterizing the molecular integrity (molar mass, state of aggregation/ degradation, conformation, flexibility and volume) of carbohydrate polymers and assemblies thereof. There are many aspects we haven't covered – mucoadhesive assemblies for example but hopefully the impression has been given that the methods considered are complementary and are particular powerful when combined together in a study, with each method providing different information on the one hand, but internal checks on the other. The following chapter by Morris *et al.* illustrates application of some of the methods described here to the study of pectins assemblies in the context of their use for efficient delivery of drugs.

Acknowledgements

The support and advice of Professor A.J. Rowe, Dr. G.G. Adams and Dr. S. Kök is appreciated. I am also grateful to T. Erten, S. Azeem, T.M.D. Besong, Dr. B. Wolf and Dr. T.J. Foster for allowing the use of data prior to publication.

References

1   S.E. Harding, *Prog. Biophys. Mol. Biol.* 1997, **68**, 207.
2   O.F. Solomon and I.Z. Ciuta, *J. Appl. Polym. Sci.* **6**, 683
3   Y. Lu, *Solution Conformation of Engineered Antibodies, PhD Dissertation*, University of Nottingham, 2007, p93.
4   E.R. Morris, A.N. Cutler, S.B. Ross-Murphy, D.A. Rees and J. Price, *Carbohyd. Polym.*, 1981, **1**, 5.
5   F. Launay, M. Milas and M. Rinaudo, *Polym. Bull.* 1982, **7**, 185.
6   C. Tanford, *Physical Chemistry of Macromolecules*, John Wiley and Sons, NewYork, 1961 p. 343.
7   O. Smidsrød and A. Haug, *Acta Chem. Scand.*, 1968, **22**, 797. See also G.M. Pavlov, Progr. Coll. Polym. Sci., 2002, **119**, 149.
8   O. Smidsrød and I.L. Andresen, *Biopolymerkjemi*, Tapir, Trondheim, 1979.
9   S.E. Harding, K. Vårum, B.T. Støkke and O. Smidsrød, *Adv. Carbohyd. Analysis*, **1**, 63.
10   D.N. Holcomb and K.E. Van Holde, *J. Phys. Chem.*, 1962, **66**, 1999.
11   P. He, S.S. Davis and L. Illum, *Int. J. Pharm.*, 1998, **166**, 75.
12   M. Fee, *Evaluation of Chitosan Stability in Aqueous Systems, PhD Dissertation*, University of Nottingham, 2005.
13   Q.E.Y. Alzahrani, *Hydrodynamic Study on Two Polysaccharide Related Substances, MSc Dissertation*, University of Nottingham, 2012.
14   J.M. Creeth, *Some Physico-chemical Studies on Nucleic Acids and Related Substances, PhD Dissertation*, University College Nottingham, 1947
15   J.M. Creeth, J.M. Gulland and D.O. Jordan, *J. Chem. Soc.*, 1947, 1141.
16   S.E. Harding and D.J. Winzor, *Macromol. Biosci.,* 2010, **10**, 696.
17   S.E. Harding, *The Independent*, 2010, 31st March, p39.
18   J.D. Watson, *The Double Helix – Illustrated and Annotated edition*, ed. A. Gann and J. Witkowski, Simon and Schuster, New York, 2012.
19   C.T. Greenwood and C. MacKenzie, *Die Stärke*, 1963, **15**, 444.
20   T.D. Bradley and J.R. Mitchell, *Carbohyd. Polym.*, 1988, **9**, 257.
21   S.E. Harding, D.B. Sattelle and V.A. Bloomfield, ed., *Laser Light Scattering in Biochemistry*, Royal Society of Chemistry, Cambridge, 1992
22   P.N. Pusey, in *Photon Correlation and Light Beating Spectroscopy*, ed. H.Z. Cummings and E.R. Pike, Plenum Press, New York, 1974, p387.
23   P. N. Pusey, in *Dynamic Properties of Biomolecular Assemblies*, ed. S.E. Harding and A.J. Rowe, Royal Society of Chemistry, Cambridge, 1989, p.?.
24   S.W. Provencher, in *Laser Light Scattering in Biochemistry*, ed. S.E. Harding, D.B. Sattelle and V.A. Bloomfield, Royal Society of Chemistry, Cambridge, 1992, p92.
25   W. Burchard in *Laser Light Scattering in Biochemistry*, ed. S.E. Harding, D.B. Sattelle and V.A. Bloomfield, Royal Society of Chemistry, Cambridge, 1992, p3.
26   T. Erten, G. G. Adams, T.J. Foster and S.E. Harding, mss. in preparation, 2012
27   S.E. Harding and P. Johnson, *Biochem. J.* 1985, **231**, 543.
28   S.E. Harding and P. Johnson, *Biochem. J.* 1985, **231**, 549.

29   P. Claes, M. Dunford, A. Kenney and P. Vardy, in *Laser Light Scattering in Biochemistry*, ed. S.E. Harding, D.B. Sattelle and V.A. Bloomfield, Royal Society of Chemistry, Cambridge, 1992, p66.

30   P.J. Wyatt, in *Laser Light Scattering in Biochemistry*, ed. S.E. Harding, D.B. Sattelle and V.A. Bloomfield, Royal Society of Chemistry, Cambridge, 1992, p35.

31   J.C. Horton, S.E. Harding & J.R. Mitchell, *Biochem. Soc. Trans.,* 1991, **19**, 510.

32   J.E. Rollings, in *Laser Light Scattering in Biochemistry*, ed. S.E. Harding, D.B. Sattelle and V.A. Bloomfield, Royal Society of Chemistry, Cambridge, 1992, p275.

33   K. Jumel, I. Fiebrig and S.E. Harding, *Int. J. Biol. Macromol.* 1996, 18, 133.

34   K. Jumel, F.J.J. Fogg, D.A. Hutton, J.P. Pearson, A. Allen and S.E. Harding, *Eur. Biophys. J.*, 1997, **25**, 477.

35   K. Jumel, S.E. Harding and J.R. Mitchell, *Carbohyd. Res.*, 1996, **282**, 223.

36   D.R. Picout, S.B. Ross Murphy, N. Errington and S.E. Harding, *Biomacromol.,* 2001, **2**, 1301.

37   T.R. Patel, D.R. Picout, S.E. Harding and S.B. Ross-Murphy *Biomacromol.*, 2006, **7**, 3513.

38   G.A. Morris, T.R. Patel, D.R. Picout, S.B. Ross-Murphy, A. Ortega, J. Garcia de la Torre and S.E. Harding, *Carbohydrate Polymers*, 2008, **72**, 356.

39   S. Azeem, T.M.D. Besong, G.G. Adams, B. Wolf and S.E. Harding, mss in preparation, 2012.

40   P. Roger, B. Baud and P. Colonna, *J. Chromatog. A*, 2001, **917**, 179.

41   S.E. Harding, A.J. Rowe and J.C. Horton, ed., *Analytical Ultracentrifugation in Biochemistry and Polymer Science*, Royal Soc. Chem.,1992.

42   S.E. Harding, in *Advances in Polymer Science. Polysaccharides I: Structure, Characterisation and Use*, ed. T. Heinze, 2005, **186**, ch. 5.

43   S.E. Harding, in S.E. Harding. A.J. Rowe and J.C. Horton ed. *Analytical Ultracentrifugation in Biochemistry and Polymer Science*, Royal Society of Chemistry, Cambridge, 1992, p.495.

44   J.M. Creeth and S.E. Harding, *J. Biochem. Biophys. Methods*, 1982, 7, 25.

45   Nikolasjski, M., Gillis, R., Adams, G.G., Berlin, P., Rowe, A.J., Heinze, T. and Harding, S.E. (mss. in preparation).

46   S.E. Harding, K. Day, R. Dhami and P.M. Lowe. *Carbohyd. Polym.* 1997, **32**, 81

47   S.E. Harding, in *Analytical Ultracentrifugation. Techniques and Methods*, ed. D. Scott, S.E. Harding and A.J. Rowe, Royal Society of Chemistry, Cambridge, 2005, p. 231.

48   M. Majzoobi, *The Effect of Processing on the Molecular Structure of Wheat Starch, PhD Thesis,* University of Nottingham, 2004.

49   R.F. Tester, T. Patel, and S.E. Harding, Carbohyd. Res., 2006, **341**, 130.

50   T. Heinze, M. Nikolajski, S. Daus, T.M.D. Besong, N. Michaelis, P. Berlin, G.A. Morris, A.J. Rowe and S.E. Harding, S. E. *Angew Chem Int Ed.,* 2011, **50**, 8602.

51   P. Schuck, Biophys. J., 2000, **78**, 1606.

52   T. Svedberg and O. Pedersen, *The Ultracentrifuge*, Oxford University Press, 1940.

53   P.N. Pusey, in *Photon Correlation and Light Beating Spectroscopy*, ed. H.Z. Cummins and E.R. Pike, Plenum Press, New York, p387.

54   P.N. Pusey, in *Dynamic Properties of Biomolecular Assemblies*, ed. S.E. Harding, and A.J. Rowe, Royal Society of Chemistry, Cambridge, 1989, p90.

55   T.R. Patel, G.A. Morris, A. Ebringerová, M. Vodenicarová, V. Velebny, A. Ortega, J. Garcia de la Torre and S.E. Harding, *Carbohyd. Polym.,* 2008, **74**, 845.

56   A. Ortega and J. Garcia de la Torre, *Biomacromol.* 2007, **8**, 2464.

57   Fujita, H. *Mathematical Theory of Sedimentation Analysis*, Academic Press, New York, 1962.
58   S.E. Harding, *Adv. Carbohyd. Chem. Biochem.*, 1989, **47**, 345.
59   S.E. Harding, P. Schuck, P., A.S. Abdelhameed, G. Adams, M.S. Kok and G.A. Morris, *Methods*, 2011, **54**, 136.
60   R.B. Gillis, G.G. Adams, B. Wolf, M. Berry, T.M.D. Besong, T.M.D., A. Corfield, S.M. Kök, R. Sidebottom, D. Lafond, A.J. Rowe, and S.E. Harding, *Carbohyd. Polym.*, 2012, in press.
61   P. Schuck, R. Gillis, G. Adams, A.J. Rowe and S.E. Harding, mss. in preparation.

G.A. Morris[1*], G.G. Adams[2], S.E. Harding[2], J.D. Castile[3] and A. Smith[3]

[1]Department of Chemical & Biological Sciences, School of Applied Sciences, University of Huddersfield, Queensgate, Huddersfield, HD1 3DH, UK
[2]National Centre for Macromolecular Hydrodynamics, School of Biosciences, University of Nottingham, Sutton Bonington, LE12 5RD, UK
[3]Archimedes Development Limited, Albert Einstein Centre, Nottingham Science Park, Nottingham, NG7 2TN, UK
*G.Morris@hud.ac.uk

1 INTRODUCTION

The delivery of drugs through the alimentary tract following oral administration via the mouth (the "oral delivery route") remains the most popular method of drug administration for both medical staff and patients alike largely due to convenience, capacity and ease of administration. However, oral drug delivery can be inefficient due to poor absorption, hostile conditions e.g. due to acidic conditions present in the stomach or degradation by digestive enzymes and particularly due to metabolism in the liver (widely known as the "hepatic first pass effect"). A relatively slow onset of action can also be an issue following oral administration even for well absorbed drugs. Recent research has therefore focused on drug absorption through nasal epithelia, although rapid clearance can be an issue for simple nasal formulations. To address these problems there has been significant interest in macromolecular-based carrier and mucoadhesive systems; these approaches are of growing importance. An outstanding example, one particularly successful system of delivery using pectin – a dietary fibre polysaccharide from the cell walls of fruit - is now utilised (PecSys®) in a nasal product recently approved for commercial sale (Fentanyl Nasal Spray marketed as PecFent® / Lazanda® by Archimedes Pharma).

In this chapter we review some of the recent work that has been conducted to characterise the physical properties of pectins that have been used for such purposes, focusing in particular on aspects of stability as reflected in the hydrodynamic and other physico-chemical properties.

2 PHYSICO-CHEMICAL PROPERTIES

Pectins are a complex family of heteropolysaccharides that constitute a large proportion of plant primary cell walls.[1-3] Pectins are made of several structural elements the most important of which are the homogalacturonan (HG) and type I rhamnogalacturonan (RG-I)

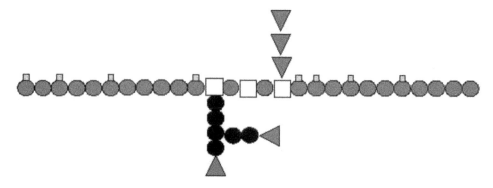

Figure 1 *Schematic structure for pectin: galacturonic acid (◐); galactose (●); arabinose (▽); rhamnose (▢) and methyl groups (▫)*

regions often described in simplified terms as the "smooth" and "hairy" regions respectively[4] (Figure 1). The HG region is composed of $(1\rightarrow4)$ linked α-D-GalpA residues that can be partially methylated at C-6 and possibly partially acetyl-esterified at O-2 and/or O-3.[5,6] Pectins with a degree of esterification (DM) > 50% are known as high methoxyl (HM) pectins and consequently low methoxyl (LM) pectins have a DM < 50%. The RG-I region consists of disaccharide repeating unit $[\rightarrow4)$-α-D-GalpA-$(1\rightarrow2)$-α-L-Rhap-$(1\rightarrow]$n with a variety side chains consisting of L-arbinosyl and D-galactosyl residues.[7] It has been reported that GalA residues in the RG-I region are partially acetylated[8,9] but not methylated.[9,10]

The degree of esterification and therefore the charge on a pectin molecule is important to the functional properties in the plant cell wall, as considered in detail in the chapter earlier in this volume by G. Tucker. It also significantly affects their commercial use as gelling and thickening agents. In some ways this is analogous to the critical effect the degree of acetylation (or deaceytylation) has on the properties of chitosans, as considered in the following chapter by Schütz, Käuper and Wandrey.

HM pectins (lower charge) form gels at low pH (< 4.0) and in the presence of a high amounts (> 55 %) of soluble solids, usually sucrose and are stabilised by hydrogen-bonding and hydrophobic interactions.[11] Conversely, LM pectins (higher charge) form electrostatically stabilised gel networks with divalent metal cations, usually calcium, in the so-called "egg-box" model, which also depends on the distribution of negative carboxylate groups and structure breaking rhamnose side chains[12,13], analogous to the alginate systems, again considered by Schütz, Käuper and Wandrey.

Solution properties such as viscosity also depend on degree of esterification, molar mass, conformation, solvent environment (e.g. salt concentration, sugar concentration and pH) and temperature (Tables 1 and 2). Hydrodynamic studies based on intrinsic viscosity ($[\eta]$), sedimentation coefficient ($s^{o}_{20,w}$), radius of gyration (R_g) and weight average molar mass (M_w) have suggested a semi-flexible conformation (Figure 2) for pectins irrespective of degree of esterification (and charge).[14-27] Pectin molar mass and chain flexibility are also important in mucoadhesive interactions.[28]

Table 1 *Representative physical properties of pectin in dilute solution*[14-27]

Physical Property	Values
Weight-average molar mass, M_w (g/mol)	30000 - 350000
Intrinsic viscosity, $[\eta]$ (ml/g)	60 - 1000
Sedimentation coefficient, $s^o_{20,w}$ (S)	1.6 – 2.3
Radius of gyration, R_g (nm)	13 - 45

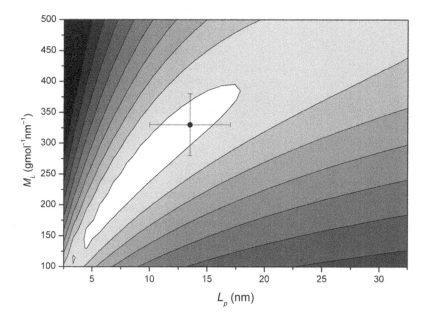

Figure 2 *Combined analysis method – HYDFIT – for pectin which estimates the best values (or best range of values of persistence length L_p and mass per unit length M_L) based on minimisation of a target function Δ. The x-axis and y-axis represent L_p (nm) and M_L (g.mol^{-1}.nm^{-1}) respectively. In these representations, the values of Δ are represented in grey shading from black to white, where for white ($\Delta = 0.2$) and black ($\Delta \geq 1$). The range of the target function minimum is typical of a semi-flexible coil where $L_p/M_L \sim 0.03 – 0.05$ nm^2.mol.g^{-1} and an estimation of the minima has been indicated*[24]

Table 2 *Representative conformation parameters of pectin in dilute solution*[14-27]

Conformational Property	Representative Values	Comments
"Power law" parameters: a b c	 0.62 – 0.94 0.17 0.57	a, b and c are Mark-Houwink-Kuhn-Sakurada (MHKS) power law relations for the molar mass dependency for intrinsic viscosity, sedimentation coefficient and root-mean-square radius respectively. For spheres, non-draining coils and rigid rods respectively (cf. the previous chapter in this volume): $a \sim 0$, 0.5-0.8 and 1.8 $b \sim 0.67$, 0.4-0.5 and 0.2 $c \sim 0.33$, 0.5-0.6 and 1.0
Frictional ratio f/f_o	7 – 10	The translational frictional ratio, f/f_o depends on conformation and molecular expansion through hydration effects. It has a value of 1 for an anhydrous sphere and increases with solvent association and with chain asymmetry.[28]
Wales-van Holde ratio $k_s/[\eta]$	0.10 – 0.85	The Wales-van Holde ratio provides a hydration independent *estimation* of conformation and has a value of ~ 1.6 for spheres and random coils and < 1.6 for asymmetric structures, approaching a lower limit of ~ 0.1 for a rigid rod.
Conformational Zone	A/B/C/D/E	A "sedimentation conformation zoning plot"[29] involving a combination of k_s with the sedimentation coefficient s and the mass per unit length M_L enables an estimate of the "overall" solution conformation of a macromolecule in solution. Zone A: rigid rod; Zone B: rod with some degree of flebility; Zone C: semi-flexible coil; Zone D: random coil; Zone E: globular sphere.[29]
Persistence length L_p (nm)	10 - 17	The linear flexibility of a polysaccharide chain can be represented quantitatively in terms of the persistence length L_p of the equivalent worm-like chain where the L_p is defined as the average projection length along the initial direction of the polymer chain (in the limit of infinite chain length).[30] Practical limits are ~ 1 nm for random coils and ~ 200 nm for a rigid rod.

3   USE IN DRUG DELIVERY

Pectins are a major natural constituent of the diet and thus are demonstrably safe. Consequently they have been used as a gelling agent (for example) in the food industry for many years. Recently there has also been significant interest in the gelling properties of pectins in controlled drug delivery.[32-35] For example, LM pectin may be used to prolong the residence of an incorporated drug substance at mucosal surfaces and thereby modulate its rate of systemic absorption. LM pectins do not act primarily as a mucoadhesive agent or absorption enhancer, instead they may alter retention and drug release characteristics due to the chelation of calcium present in nasal secretions; this results in the formation of a gel in which pectin chains are linked in an ordered three-dimensional network cross-linked by calcium ions.[36] Pharmaceutical interest in pectins is also in part due to their long standing reputation of their GRAS (Generally Regarded As Safe) status,[33,34,37] their relatively low production costs[32] and high availability[38]. It is proposed that pectin could be used to deliver drugs orally, nasally, topically and vaginally[32,33,39-43] which are routes that are generally well accepted by patients.[33,34,44] Pectin gels are pseudoplastic[32,42,43] and drug release is diffusion-controlled at low pectin concentrations[33,42] and determined by gel dissolution at higher pectin concentration.[33] The use of pectins in modulating drug absorption is particularly noteworthy in the field of nasal drug delivery. The utility of nasal route is limited by the small sample volume of up to 200 µl that can be delivered,[45] which is important in drug formulations especially if the drug is sparingly soluble or if a drug has to be delivered over a prolonged period. Even at these low volumes run-off from the nose and/or dripping into the throat can be an issue for a simple nasal spray solution.[45,46] PecSys®, a gelling technology based on LM pectin has been developed to optimise nasal delivery by addressing problems such as this. The utility of PecSys® for enhanced nasal delivery has been demonstrated in the clinic with a number of molecules including fentanyl and buprenorphine.[37]

3.1 Case study – Fentanyl Nasal Spray

Fentanyl Nasal Spray (marketed as PecFent® [EU] and Lazanda® [US] by Archimedes Pharma) is indicated for the management of breakthrough pain in adults receiving maintenance opioid therapy for chronic cancer pain. Breakthrough pain in cancer (BTPc) is a transitory exacerbation of pain that occurs on a background of otherwise controlled persistent pain. It is an intense, sudden pain that is often unpredictable and debilitating and occurs despite otherwise appropriate opioid therapy for background pain. BTPc often has high intensity, a rapid onset, usually reaching maximum intensity within five minutes, and a short duration, lasting between 30 to 60 minutes per episode. On average, BTPc affects more than half of patients with cancer and often interferes with patients' health and ability to engage in daily living activities. Given the profile of such a BTPc episode, onset of action is of primary importance in managing the condition. Fentanyl is the gold standard opioid for management of BTPc since it acts rapidly once it enters the systemic circulation and its duration of action closely matches the duration of a typical pain episode.

In addition to modulating absorption to provide rapid but controlled delivery of fentanyl, [37,47,48] PecSys® technology is employed to help avoid run-off from the nose and/or dripping into the throat[45] that can be associated with conventional nasal spray formulations and thus further optimise efficiency of drug absorption. The LM pectin element of PecSys® causes the product to gel *in situ* when sprayed into the nose due to the interaction of LM pectin with calcium in endogenous nasal secretions. *In situ* gelling ensures that the product can be sprayed readily as a finely dispersed plume before forming a gel on contact with the mucosal surface.

4 STABILITY

Pectins are not only subject to enzymatic degradation – as we have seen already in the previous chapters in this volume by Tucker and Grassby *et al.* – but also to thermal and other degradative effects of bioprocessing. In solution, depending on conditions, they can degrade by de-esterification and depolymerisation. The extent and rate of degradation depends on factors such as pH, water activity and temperature[45], so choice of the optimum storage conditions is essential in commercial products. In general, the maximum stability is found at pH ~ 4.[49] The presence of sugar also has a certain protective effect.[49] At low pH-values and elevated temperatures degradation is due to depolymerisation whilst de-esterification is also favoured. At pH 6, HM-pectin is stable is relatively stable at moderate temperatures (~20 °C), however as temperature or pH increases β-elimination will occur which results in depolymerisation and a loss of viscosity[23] and gelling properties.[32,50] LM-pectins are more stable with respect to depolymerisation under these conditions (Figure 3).[51] HM-pectins, as a solid form, lose their ability to form gels if stored in humid conditions whilst LM-pectins are again more stable and loss is not expected to be significant after one year storage at room temperature[49], an observation that is consistent with the use of LM pectin in pharmaceutical products that have remained demonstrably stable for up to 3 years.

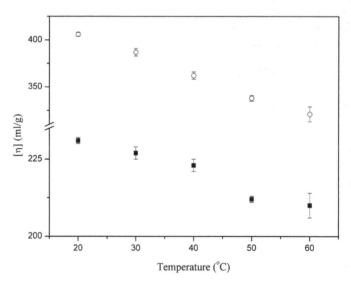

Figure 3 *The effect of increased temperature on the intrinsic viscosity, [η], for LM (■) and HM pectins (○) in phosphate-chloride buffer (pH ~ 6.8; I = 0.1 M) adapted from Figures 4 and 2 in references 50 and 23, respectively. In the case of the LM pectin, the decrease in intrinsic viscosity is a least partially due to change in conformation to a less rigid structure*

The stability (shelf-life) of pectin in terms of viscosity and gel strength is highly relevant to its commercial uses as these properties can play an important role in the function of pectin.[33,34] It is, therefore, fundamentally important to have the means available with which to measure the effects of and understand the relationships between storage conditions and stability.

We have previously looked at the stability of pectin solutions, in terms of viscosity, across a range of different temperature conditions: 4°C, 25°C and 40°C and the consequent effect on molar mass and gel strength.[52,53] The viscosity of pectin solutions decreases marginally after 6 months storage at 25 °C and significantly after 6 months at 40 °C (Table 3). A correlation between decrease in viscosity and gel strength upon addition of calcium ions has been demonstrated (Figure 4; Table 3)[53], observations which can be ascribed to a significant depolymerisation of the pectin over time (Figure 5).

Table 3 *Solution viscosities, intrinsic viscosities, gel strengths and molecular weights for pectin of degree esterification 19% (P_{19}) stored at different temperatures (4°C, 25°C or 40°C). Adapted from Table 2 in ref. 53*

Storage time (days)	Storage temperature (°C)	Viscosity (mPas)	Intrinsic viscosity (mL/g)	Weight-average molar mass (g/mol)	Gel strength area (g.sec)
0	-	4.41 ± 0.01	396 ± 1	154000 ± 1000	1645 ± 105
30	4	4.52 ± 0.07	405 ± 6	158000 ± 3000	1305 ± 200
90	4	4.48 ± 0.01	402 ± 1	156000 ± 1000	1400 ± 235
180	4	4.45 ± 0.02	399 ± 2	155000 ± 1000	1320 ± 340
30	25	4.40 ± 0.05	395 ± 4	153000 ± 2000	1340 ± 20
90	25	4.24 ± 0.02	383 ± 1	147000 ± 1000	1340 ± 380
180	25	4.05 ± 0.02	367 ± 2	140000 ± 1000	1000 ± 265
30	40	3.75 ± 0.06	341 ± 5	128000 ± 2000	935 ± 25
90	40	3.12 ± 0.01	282 ± 1	103000 ± 1000	1035 ± 25
180	40	2.45 ± 0.01	212 ± 1	73000 ± 1000	860 ± 80

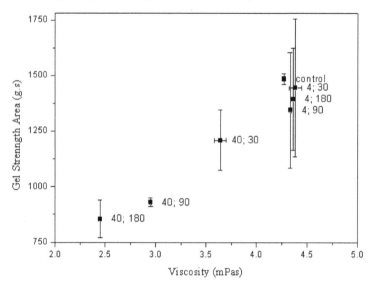

Figure 4 *The relationship between gel strength and viscosity for LM pectins of DM 21 % (labels indicate the temperature (°C); and the time (days) at which a measurement was taken)*

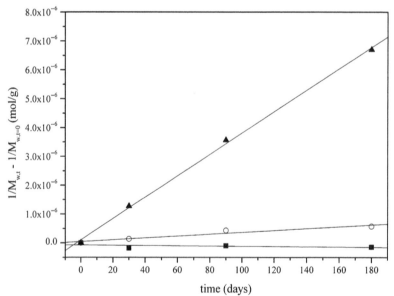

Figure 5 *First order kinetic plots of (mol/g) vs. time (days) for pectin of DM ~ 19%, where closed symbols represent molar masses estimated from viscometry at 4 °C (■), 25 °C (○) and 40 °C (▲) (adapted from Figure 3 in 35). The kinetic rate constants (day^{-1}) are (-0.8 ± 1.1) x 10^{-7}, (5.7 ± 1.1) x 10^{-7} and (6.7 ± 0.2) x 10^{-6} at 4 °C, 25 °C and 40 °C, respectively*

It has been shown that drug release rates from pectin gels *in vitro*[36,42] are relatively unaffected by a decrease in absolute viscosity from approximately 5 – 2 mPas[52,53] which suggests that a decrease in pectin molar mass from 175000 – 50000 g/mol would have no significant effect on drug release rates from pectin gels (Figure 6). This therefore implies that even at 40 °C there will be no significant effect on drug release upon storage of up to 1 year. In calcium pectate based tablet formulations drug release time is increased with lower degree of methyl esterification, but higher levels of calcium ions can lead to disintegration of the tablet and increased drug release.[32] Similarly diffusion profiles of fentanyl from LM pectin-based (PecSys[®]) nasal liquid formulations have been studied,[37] and have been shown to remain stable throughout product shelf-life.

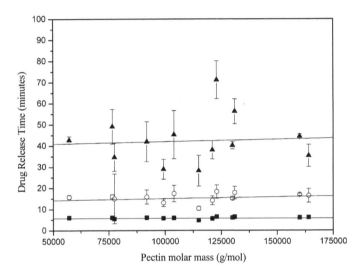

Figure 6 *Effect of pectin molar mass on model drug (paracetamol) release from pectin gel systems. 10% drug release (■), 50% drug release (○) and 90% drug release (▲)*

5 CONCLUSIONS

Over the last few decades there has been significant interest in the use of polysaccharides such as pectin in drug delivery systems, with the realisation of some successful and demonstrably stable products (*e.g.* PecFent[®] / Lazanda[®] Nasal Spray). However, in view of the polydisperse nature of the biopolymer itself it remains necessary to evaluate, refine and optimise any pectin-based drugs delivery systems for improvement and refinement for a particular application with respect to issues such as optimal molar mass and molar mass distribution, macromolecular conformation[54], degree of esterification and distribution of methyl groups[4]. Further appreciation of these parameters will provide a better understanding of molecular interactions[55] and improved ways of optimising stability (in terms of molar mass/viscosity/gel strength) and the effect of temperature and pH.[23,35,50] Another fruitful area of further study would appear to be an exploration of interactions with other polysaccharides *e.g.* cationically charged chitosan[55] and the effect of mechanical[56], chemical[56,57] or enzymatic[2,58] modification on the any of the above issues.

Acknowledgements

We would like to thank Dr. M-C. Ralet, Dr. T. J. Foster, Mrs. M-J. Crépeau and Mrs. S. Daniel for their expertise and we would also like to thank both the Région Pays de la Loire and the United Kingdom Biotechnology and Biological Sciences Research Council (BBSRC) for their financial support (BB/D01364X/1).

References

1   W. G. T. Willats, L. McCartney, W. Mackie and J. P. Knox, *Plant Mol. Biol.*, 2001, **47**, 9.
2   G. A. Morris, M-C. Ralet, E. Bonnin, J-F. Thibault and S. E. Harding, *Carbohyd. Polym.*, 2010, **82**, 1161.
3   G. A. Morris and M-C. Ralet, *Carbohyd. Polym.*, 2012, **87**, 1139.
4   M. A. K. Williams, A. Cucheval, A. Ström and M-C. Ralet, *Biomacromolecules*, 2009, **10**, 1523.
5   W. Pilnik and A. G. J. Voragen in *Biochemistry of Fruits and their Products*, ed. A. C. Hulme, Academic Press, New York, 1970, 53.
6   F. M. Rombouts and J-F. Thibault in *The Chemistry and Function of Pectins,* ed. M. L. Fishman and J. J. Jen, American Chemical Society, Washington DC, 1986, 49.
7   A. G. J. Voragen, W. Pilnik, J-F. Thibault, M. A. V. Axelos and C. M. G. C. Renard in *Food Polysaccharides and their Interactions* ed. A. M. Stephen, Marcel Dekker, New York, 1995, 287.
8   T. Ishii, *Plant Physiol.*, 1997, **113**, 1265.
9   P. Perrone, H. C. Hewage, H. R. Thomson, K. Bailey, I. H. Sadler and S. C. Fry, *Phytochemistry*, 2002, **60**, 67.
10   P. Komalavilas and A. J. Mort, *Carbohyd. Res.*, 1989,**189**, 261.
11   J. A. Lopes da Silva and M. P. Gonçalves, *Carbohyd. Polym.*, 1994, **24**, 235.
12   E. R. Morris, *Food Chem.*, 1980, **6**, 15.
13   E. R. Morris, D. A. Powell, M. J. Gidley and D. A. Rees, *J. Mol. Biol.*, 1982, **155**, 507.
14   H. Anger and G. Berth, *Carbohyd. Polym.*, 1985, **5**, 241.
15   M. A. V. Axelos, J. Lefebvre and J-F. Thibault, *Food Hydrocolloid.*, **1**, 1987, 569.
16   G. Berth, H. Anger and F. Linow, *Nahrung-Food*, 1977, **31**, 939.
17   S. E. Harding, G. Berth, A. Ball, J. R. Mitchell and J. Garcìa de la Torre, *Carbohyd. Polym.*, 1991, **16**, 1.
18   C. Garnier, M. A. V. Axelos and J-F. Thibault, *Carbohyd. Res.*, 1993, **240**, 219.
19   A. Malovikova, M. Rinaudo and M. Milas, *Carbohyd. Polym.*, 1993, **22**, 87.
20   S. C. Cros, C. Garnier, M. A. V. Axelos, A. Imbery and S. Perez, *Biopolymers*, 1996, **39**, 339.
21   I. Braccini, R. P. Grasso and S. Perez, *Carbohyd. Res.*, 1999, **317**, 119.
22   G. A. Morris, T. J. Foster and S. E. Harding, *Food Hydrocolloid.*, 2000, **14**, 227.
23   G. A. Morris, T. J. Foster and S. E. Harding, *Carbohyd. Polym.*, 2002, **48**, 361.
24   G. A. Morris, J. García de la Torre, A. Ortega, J. Castile, A. Smith and S. E. Harding, *Food Hydrocolloid.*, 2008, **22**, 1435.
25   M. L. Fishman, H. K. Chau, P. D. Hoagland and A. T. Hotchkiss, *Food Hydrocolloid.*, 2006, **20**, 1170.
26   M. L. Fishman, H. K. Chau, F. Kolpak and J. Brady, *J. Agr. Food Chem.*, 2001, **49**, 4494.
27   R. Noto, V. Martorana, D. Bulone and B. L. San Biagio, *Biomacromolecules*, 2005, **6**, 2555.

28   C. Tanford. *Physical Chemistry of Macromolecules*. John Wiley and Sons, New York, 1961.

29   G. M. Pavlov, A. J. Rowe, and S. E. Harding, *Trends Anal. Chem.*, 1997, **16**, 401.

30   O. Kratky, and G. Porod, *Recl. Trav. Chim. Pay-B.*, 1949, **68**, 1106.

31   N. A. Nafee, F. A. Ismail, N. A. Boraie and L. M. Mortada, *Drug Dev. Ind. Pharm.*, 2004, **30**, 985.

32   S. Sungthongjeen, P. Sriamornsak, T. Pitaksuteepong, A. Somsiri and S. Puttipipatkhachorn, *AAPS PharmSciTech*, 2004, **5**, 1.

33   L. Lui, M. L. Fishman and K. B. Hicks, *Cellulose*, 2007, **14**, 15.

34   L. Lui, M. L. Fishman, J. Kost and K. B. Hicks, *Biomaterials*, 2003, **24**, 3333.

35   G. A. Morris, J. Castile, A. Smith, G. G. Adams and S. E. Harding, *Polym. Degrad. Stabil.*, 2010, **94**, 1344.

36   S. T. Charlton, S. S. Davis and L. Illum, *J. Control. Release*, 2007, **118**, 225.

37   P. Watts and A. Smith, *Expert Opin. Drug Del.*, 2009, **6**, 543.

38   C. M. Beneke, A. M. Viljoen and J. H. Hamman, *Molecules*, 2009, **14**, 2602.

39   N. A. Peppas, P. Bures, W. Leobandung and H. Ichikawa, *Eur. J. Pharm. Biopharm.*, 2000, **50**, 27.

40   V. R. Sinha and R. Kumria, *Int. J. Pharm.*, 2001, **224**, 19.

41   C. Valenta, *Adv. Drug Deliver. Rev*, 2005, **57**, 1692.

42   S. Chelladurai, M. Mishra and B. Mishra, *Chem. Pharm. Bull.*, 2008, **56**, 1596.

43   N. Thirawong, R. A. Kennedy and P. Sriamornsak, *Carbohyd. Polym.*, 2008, **71**, 170.

44   N. Yadav, G. A. Morris, S. E., Harding, S. Ang and G. G. Adams, *Endocr., Metab. Immune Disord.: Drug Targets*, 2009, **9**, 1.

45   J. Castile, Y.-H. Cheng, B. Simmons, M. Perelman, A. Smith and P. Watts, *Drug Dev. Ind. Pharm.* 2012, (in press)

46   E. O. Meltzer, J. E. Stahlman, J. Leflein, S. Meltzer, J. Lim , A. A. Dalal, B. A. Prillaman and E. E. Philpot, *Clin. Therap.* 2008, **30**, 271.

47   A. N. Fisher, M. Watling, A. Smith and A. Knight, *Int. J. Clin. Pharmacol. Ther.*, 2010, **48**, 138.

48   A. N. Fisher, M. Watling, A. Smith and A. Knight, *Int. J. Clin. Pharmacol. Ther.* 2010, **48**, 860.

49   P. Sriamornsak, *Silpakorn University International Journal*, 2003, **3**, 206.

50   G. A. Morris, T. J. Foster and S. E. Harding, *Prog. Coll. Pol. Sci.,* 1999, **113**, 205.

51   C. Rolin, In *Industrial Gums*, ed. R. L. Whistler and J. N. BeMiller, Academic Press, New York, 1993, 257.

52   G. A. Morris, M. S. Kök, S. E. Harding and G. G. Adams, *Biotechnol. Genet. Eng. Rev.*, 2010, **27**, 257.

53   G. A. Morris, J. D. Castile, A. Smith, G. G. Adams and S. E. Harding, *Polym. Degrad. Stabil.,* 2011, **95**, 2670.

54   G. A. Morris and M-C. Ralet, *Carbohyd. Polym.*, 2012, **88**, 1488.

55   S. E. Harding, *Trends Food Sci. Tech.* 2006, **17**, 255.

56   U. Einhorn-Stoll, H. Kunzek, and G. Dongowski, *Food Hydrocolloid.*, 2007, **21**, 1101.

57   G. A. Morris, Z. Hromadkova, A. Ebringerova, A. Malikova, J. Alfoldi and S. E. Harding, *Carbohyd. Polym.*, 2002, **48**, 351.

58   K. T. Inngjerdingen, T. Patel, X. Chen, L. Kenne, S. Allen, G. A. Morris, S. E. Harding, T. Matsumoto, D. Diallo, H. Yamada, T. E. Michaelsen, M. Inngjerdingen and B. S. Paulsen, *Glycobiology*, 2007, **17**, 1299.

9

STABILITY OF POLYSACCHARIDE COMPLEXES: THE EFFECT OF MEDIA AND TEMPERATURE ON THE PHYSICAL CHARACTERISTICS AND STABILITY OF CHITOSAN-TRIPHOSPHATE/ ALGINATE NANOGELS

C.A. Schütz[1], P. Käuper[2] and C. Wandrey[1*]

[1]Institut d'Ingénierie Biologique et Institut des Sciences et Ingénierie Chimiques, Ecole Polytechnique Fédérale de Lausanne, CH-1015 Lausanne, Switzerland
[2]Medipol SA, PSE-B, CH-1015 Lausanne, Switzerland
*christine.wandrey@epfl.ch

1 INTRODUCTION

Ionic polysaccharides are able to form complexes by electrostatic interactions. Such complexes are becoming increasingly interesting for biomedical and pharmaceutical applications. A variety of complexes is under investigation as scaffolds, to immobilize/encapsulate cells or tissue, to deliver bioactive materials to specific sites, or even to enter cells. Of special interest are micro- and nano-sized spherical materials, which retain a large quantity of water, and which are therefore classified as *hydrogels*. Chitosan and sodium alginate (Na-alg) belong to the preferred polysaccharide candidates. Hydrophilic micro- and nanocarriers formed from these are considered as having an enormous potential for biomedical and pharmaceutical applications. Complexes containing chitosan and Na-alg as components are also the subject of this contribution.

Chitosan is a modified natural polysaccharide derived from chitin by deacetylation. It is the second most abundant polysaccharide in nature after cellulose.[1] The major source of chitin is the exoskeleton of crustaceans. However, it is also present in the cell wall of fungi. Ultrapure chitosan manufactured from an edible fungal source, *Agaricus bisporus*, free of animal byproducts, is an interesting alternative to replace animal chitosan in biomedical and pharmaceutical applications.[2] The cationic character of chitosan in solution allows for electrostatic interaction with negatively charged multivalent ions, small molecules, or polyanions. Applying specific technologies, nanoparticulate complexes can be formed. Such chitosan-based nanoparticles have been proposed for a variety of applications including drug delivery,[3] DNA and RNA delivery,[4,5] and vaccine nanoparticulate systems.[6]

Preferred polyanions for forming hydrophilic nano-sized chitosan complexes are pentasodium triphosphate (TPP) and Na-alg. Due to the high water content of the physically cross-linked polymer network of the nanoparticles, frequently the term *nanogels* is used. Na-alg is a linear copolymer composed of mannuronate and guluronate units. Commercial varieties of alginate are extracted from seaweed. Due to its widely demonstrated biocompatibility and simple gelation - in particular with divalent cations -

this anionic biopolymer is under scrutiny for a multitude of biomedical applications.[7-10] The quality of the complexes with chitosan strongly depends on the medium conditions, in particular the pH.[11]

The knowledge of nanogel characteristics such as size, size distribution, surface charge, and stability under a variety of conditions is crucial for the development and application of efficient nanoparticulate therapeutic systems. The physical characteristics depend on the formation technology as well as storage and application environment, on parameters such as temperature, pH, ionic strength, or specific ionic and non-ionic additives. Recently, several authors have reported the storage behavior and stability of chitosan/TPP nanoparticles.[12,13] Tsai *et al.* have published the size and stability of positively charged chitosan-TPP complexes in phosphate buffer over time.[13] The results showed the existence of three stages of particle size variation, an initial instantaneous stage of size reduction, a stable ageing period followed by a swelling/aggregation stage after 10 days. However, whether the study was performed under sterile conditions was not obvious. Natural polysaccharides, chitosan and Na-alg solutions are particularly subject to microorganism contamination, which could strongly influence the characteristics and stability of the nanoparticles during storage. Morris *et al.* have studied the particle size of chitosan/TPP complexes in acetate buffer at temperatures between 4°C and 40°C as a function of time.[12] Their study revealed decrease of the particle size for storage at 40°C after 1 month and complete disintegration after 6 months. This observation was attributed to decreasing molar mass of chitosan. The particle size of the chitosan/TPP complexes stored at 4°C and 25°C remained stable.

Special attention is needed concerning the frequent difference between preparation and storage conditions on the one hand and application conditions on the other. This concerns mainly the quality of the medium but also the temperature. Knowledge of these effects is crucial for the performance of nanogels intended for applications in a biological environment.

The objective of the study presented in this Chapter was to characterize spherical nanogels, which had been produced by ionotropic gelation of chitosan with pentasodium triphosphate (TPP) followed by coating with Na-alg in different media and at different temperatures, and then to compare their physical properties as obtained in the production medium water with those approached in PBS and different biologically relevant liquids. Chitosan obtained from both animal and fungal sources was considered. Of particular interest was the stability in both the media and the temperatures which are relevant for cell survival and at storage conditions with respect to biomedical or pharmaceutical applications.

Previously, these nanogels have been examined as delivery vehicles *in vitro* and evaluated concerning their cytocompatibility.[14] Knowledge of the stability will contribute to further progress in the development of these nanogels, in particular for helping us discover or confirm the mechanisms of cell uptake and intracellular fate.

2 METHODS

2.1 Chitosan Purification

Chitosan from crustacean shell (Primex, Iceland) and chitosan biotechnologically produced from *Agaricus bisporus* under GMP (Good Manufacturing Practice) conditions (Kitozyme, Herstal, Belgium) were used. Chitosan was purified prior to analysis as described previously.[15] The procedure included dissolution at 1% w/v in HCl approx. pH 4, sterile filtration, precipitation by adding 1M NaOH dropwise under strong agitation, dialysis, and freeze-drying.

2.2 Chitosan Characterization

Chitosan from animal and fungal sources was characterized by ^{1}H NMR spectroscopy, dilution viscometry and analytical ultracentrifugation. The degree of deacetylation (DDA) was analyzed by ^{1}H NMR spectroscopy according to peak assignment and DDA determination reported in the literature.[16-18] To determine the intrinsic viscosity [η], chitosan was dissolved in a solution of 0.02 M acetate buffer and 0.2 M NaCl, and the intrinsic viscosity was extrapolated according to Huggins.[19]

Sedimentation velocity experiments were performed in an analytical ultracentrifuge OPTIMA XL-I/ProteomLab™ (Beckman Coulter, Palo Alto, USA) using the Rayleigh interference optics. The samples were run at three concentrations in the range of 1.2 to 1.9 mg/mL. A 4-hole titanium rotor was used equipped with three double-sector titanium cells and sapphire windows (Beckman Coulter, Palo Alto, USA). Runs were carried out at 20°C with 400 μL solvent and 380 μL sample volume and a rotor speed of 5.5×10^4 rpm. Two hundred scans were collected at 5 min intervals. Sedimentation coefficients were calculated using the *SEDFIT* software.[20] The apparent sedimentation coefficients (s_{app}) were plotted vs. the chitosan concentration to obtain s^o_{20} and the parameter k_s according to Equation 1:

$$\frac{1}{s_{app}} = \frac{1}{s^o_{20}} + \left(k_s \cdot \frac{1}{s^o_{20}} \right) c \tag{1}$$

2.3 Nanogel Formation

Figure 1 demonstrates the nanogel formation based on a previously developed and described process.[14,21]

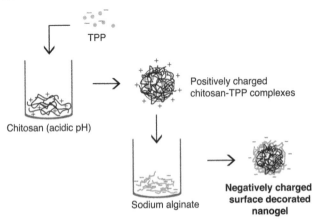

Figure 1 *Nanogel formation by ionic gelation of chitosan pentasodium triphosphate (TPP) and subsequent coating with sodium alginate*

Two types of nanogels were prepared, one empty and the other loaded with a model protein. Under sterile conditions, a 0.1% w/v sterile filtered TPP solution containing 1 mg/mL bovine serum albumin (BSA) in the case of the loaded nanogels, was added drop-wise to 9-fold volume of a sterile filtered 0.1% w/v chitosan solution (pH 3.5 - 4.0) under agitation. For the nanogel surface coating, the formulation was diluted 1:1 with sterile water, and added drop-wise to an equal volume of 0.1% w/v sterile filtered aqueous Na-alg solution under agitation maintaining the pH between 7.1 and 7.5. The dispersion was filtered through a 1.2 μm hydrophilic filter (Minisart, Sartorius).

Na-alg, a purified batch of Keltone LVCR (ISP, San Diego, USA) with a mannuronic/guluronic acid ratio of approx. 60/40, was provided by Medipol (Medipol SA, Switzerland). TPP purum p.a. ≥ 98% TPP and BSA were purchased from Sigma-Aldrich (Sigma-Aldrich, Germany).

2.5 Nanogel Characterization

The size distribution and the surface charge of the nanogels were analyzed concerning the influence of temperature, mechanical stress, medium composition, as well as long-term stability. A Malvern ZetaSizer (ZEN3600 Nano-ZS, Malvern Instruments, UK) was used. The setting parameters used in the standard operation procedure (SOP) are summarized in Table 1.

Table 1 *Parameter setting of the ZetaSizer Nano-ZS for the analysis of size distribution by dynamic light scattering and zeta potential by electrophoretic mobility*

Parameter	Setting
Equilibrium time* (s)	180
Temperature** (°C)	25
Refractive index	1.59
Absorption	0.0
	0.3 *(samples in DMEM)*
Dispersant	Water
Number of measures	3
Runs	Automatic
Angle	173° - Backscatter (size)
Model	Smoluchowski (zeta)

*at the beginning of each single measurement or cycle of measurements
**if not specified differently

2.5.1 Influence of the Ionic Strength and pH. Subsequent to the preparation of the nanogels in water, they were transferred into phosphate buffered saline (PBS) and Dubelcco's minimum essential medium (DMEM). DMEM powder was purchased from Sigma-Aldrich (Sigma-Aldrich, Germany). PBS 10x was obtained from internal stocks (Pharmacy, CHUV, Switzerland). A 7x concentrated DMEM solution was prepared by dissolving 1 L powder equivalent in 143 mL sterile milliQ water, adding sodium bicarbonate (3.7 g), adjusting the pH if necessary to 7.4. The concentrated medium was sterile filtered through filters of 0.22 μm.

2.5.2 Influence of the Medium Temperature. To investigate the influence of the medium temperature on the nanogel characteristics, the "SOP player" function of the ZetaSizer was used to automatically changing the temperature of the nanogel suspension between 4°C and 37°C over time and acquire data at each time point predetermined. The temperature was automatically equilibrated inside the cuvette chamber at each temperature selection. The nanogels were taken directly from 4°C storage and placed in the sample chamber priorly equilibrated at 4°C.

2.5.3 Influence of Biologically Relevant Media. To simulate *in vivo* conditions, nanogels were exposed to three different media: DMEM with 10% v/v heat-inactivated fetal bovine serum (FBS) from cell culture, DMEM with 1% w/v BSA, and cell media supernatant collected from human colon adenocarcinoma (HT-29), and human brain-derived endothelial (HCEC) cell cultures.[14]

2.5.4 Mechanical Stability. Filtration through filters of different pore size was utilized to evaluate the mechanical stability. After formation in water, nanogels were stored overnight at 4°C and then filtered sequentially through 1.2 μm, 0.8 μm, 0.45 μm, 0.2 μm and 0.1 μm hydrophilic syringe filters (Minisart, Sartorius; VWR International SA, Switzerland for the 0.45 μm filters). Only manual filtration by the same manipulator (no vacuum) was applied to filter the suspensions. Size distribution and zeta potential were determined separately for each fraction.

2.5.5 Long-term Stability. The influence of media and temperature on the durability and long-term stability of the nanogels was investigated according to the scheme presented in Figure 2. Sterile BSA-loaded nanogels were stored in 4 mL glass vials in water as prepared, or in PBS, at 4°C and at room temperature. Size distribution and zeta potential were measured at day 0 in water and subsequently in water and PBS over a period of two months. Visual observation was a first indication of instability or precipitation.

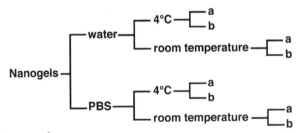

Figure 2 *Storage scheme*

3 RESULTS AND DISCUSSION

3.1 Comparison of Animal and Fungal Chitosan

Knowledge of the characteristics of the animal and fungal chitosan is crucial for the comparison of the nanogels prepared thereof. Table 2 summarizes the chitosan characteristics for both samples. Previous characterization revealed similar DDA but a higher intrinsic viscosity for the animal chitosan, 449 mL/g compared to 329 mL/g for the fungal sample.[14] The values of the Huggins coefficient k_H (0.37-0.43) confirm similar and good solubilization.

A more detailed characterization was performed by analytical ultracentrifugation and revealed similar sedimentation coefficient distributions for both the animal and the fungal chitosan (Figure 3). Via k_s determined from the concentration dependence of the sedimentation coefficients, the Wales-van Holde ratio $k_s/[\eta]$ was calculated. It is considered as a measure of the macromolecule's "overall" conformation in solution. The $k_s/[\eta]$ ratios are similar for the animal and fungal chitosan (Table 2). This confirms that despite the different intrinsic viscosity values, the conformation in solution can be concluded as being similar. The values for the $k_s/[\eta]$ ratios are in agreement with the data range for chitosan reported by Pavlov and Selyunmin.[22] The values of 0.57-0.62 obtained here suggest an extended conformation.[23,24] Several controversial results have been published concerning the conformation of chitosan molecules, concluding rod-like chains or flexible coils. Nevertheless, the consensus of defining the conformation as "semi-flexible rod" or "stiff coil" is in good agreement with the values obtained here.[25] More importantly, the values of $k_s/[\eta]$ are similar for animal and fungal chitosan samples tested here and confirm similar behavior in solution. This is crucial for further development of nanocarrier systems based on these polysaccharides. Moreover, these results are in good agreement with other polysaccharide values for $k_s/[\eta]$, such as 0.6 obtained for alginate.[26]

Table 2 *Degree of deacetylation (DDA), intrinsic viscosity [η], Huggins coefficients k_H (mean ± SD, n≥3), sedimentation coefficient s^0_{20}, constant k_s, and Wales-van Holde ratio $k_s/[\eta]$*

Chitosan	DDA (%)	$[\eta]_H$ (mL/g)	k_H	s^0_{20} (S)	k_s (mL/g)	$k_s / [\eta]$
Animal	87	449 ± 58	0.43 ± 0.10	1.80	280	0.62
Fungal	90	329 ± 18	0.37 ± 0.10	1.60	187	0.57

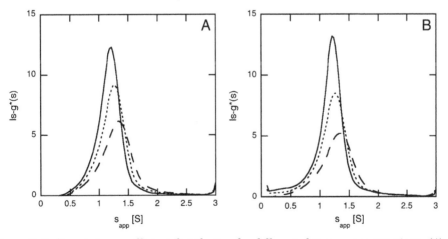

Figure 3 *Sedimentation coefficient distribution for different chitosan concentrations. (A) animal chitosan, (line) 1.9 mg/mL, (dotted) 1.6 mg/mL, (dashed) 1.3 mg/mL. (B) fungal chitosan, (line) 1.9 mg/mL, (dotted) 1.6 mg/mL, (dashed) 1.2 mg/mL*

3.2 Size, Charge and Stability of Nanogels in Different Media

Sphericity with diameters in the range of 100 to 350 nm was monitored by electron microscopy (SEM) for dried nanogels.[14] As expected, the diameters obtained by electron microscopy are significantly smaller than those analyzed in solution by dynamic light scattering (DLS). The latter show a wide distribution verified by high polydispersity index (PdI) values between 0.4-0.5 in water. The z-average of the hydrodynamic diameter, $d_{H,Z}$, and the PdI of empty and BSA-loaded nanogels produced with animal or fungal chitosan are summarized in Table 3 as mean of nanogel suspension replicates. Table 4 shows the zeta potential for the same preparations.

Nanogels prepared with fungal chitosan are smaller than those prepared with animal chitosan, although the differences are not significant. This difference can be attributed to the different intrinsic viscosity/molar mass of the chitosan samples (Table 2). However, the diameter difference, < 15%, is less pronounced than the difference estimated for the molar mass, about 50%.[14]

Table 3 *Hydrodynamic diameter as z-average ($d_{H,Z}$) and polydispersity index (PDI) of nanogels prepared from animal or fungal chitosan, empty or loaded with BSA*

		$d_{H,Z}$ (nm)		PDI	
Nanogel	Medium	Animal	Fungal	Animal	Fungal
Empty	H$_2$O	558 ± 60	480 ± 21	0.47 ± 0.05	0.42 ± 0.02
	PBS	406 ± 41	350 ± 29	0.25 ± 0.02	0.23 ± 0.00
	DMEM	294 ± 40	286 ± 54	0.24 ± 0.11	0.20 ± 0.02
BSA-loaded	H$_2$O	513 ± 0	493 ± 2	0.46 ± 0.00	0.39 ± 0.01
	PBS	387 ± 23	346 ± 17	0.27 ± 0.02	0.26 ± 0.02
	DMEM	310 ± 0	273 ± 0	0.16 ± 0.00	0.18 ± 0.00

Table 4 *Zeta potential ζ of nanogels prepared from animal or fungal chitosan, empty or loaded with BSA*

		ζ potential [mV]	
Nanogel	Medium	Animal	Fungal
Empty	H$_2$O	-64 ± 0	-66 ± 2
	PBS	-27 ± 2	-28 ± 0
	DMEM	-25 ± 1	-27 ± 1
BSA-loaded	H$_2$O	-62 ± 0	-66 ± 1
	PBS	-27 ± 1	-27 ± 0
	DMEM	-24 ± 0	-25 ± 0

For all nanogel preparations, the $d_{H,Z}$ was significantly reduced in PBS and even more in the cell culture medium DMEM, confirming the expected influence of the ionic strength and the pH on the size of such nano-sized hydrogels. The contraction of the polyelectrolyte network in high ionic strength media corresponds to the theory. The PDI values are also significantly reduced compared to water, with values between 0.2-0.25. Size distributions are well defined in PBS and DMEM (Figures 4C, 4D and 4E, 4F), in contrast to nanogels in water where a shoulder peak was observed (Figures 4A, 4B). Volume and number representations of particle size distributions (not shown) are similar to the intensity frequency distributions in Figure 4. This observation confirms the presence of only one size population of particles, although it is still a wide distribution. Loading a protein cargo in the nanogels (here BSA) did not significantly modify the $d_{H,Z}$ for neither animal nor fungal chitosan nanogels. Size distributions are similar for both empty and BSA loaded nanogels (Figure 4A, C, E vs. 4B, D, F).

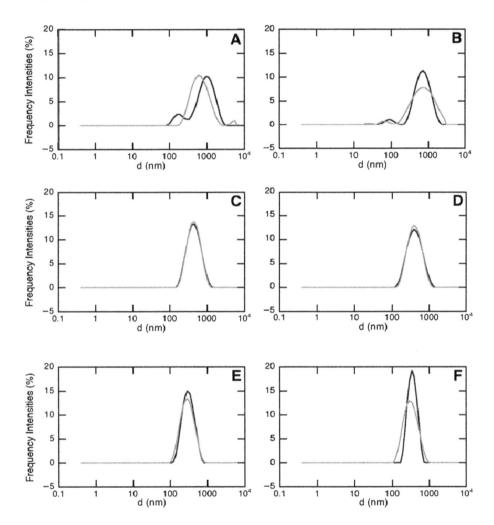

Figure 4 *Size distributions as intensities for representative samples. (A, B) as prepared in water, (C, D) in PBS, (E, F) in DMEM. (A, C, E) chitosan-TPP/alginate nanogels, (B, D F) BSA loaded chitosan-TPP/alginate nanogels, (black line) animal chitosan, (grey line) fungal chitosan*

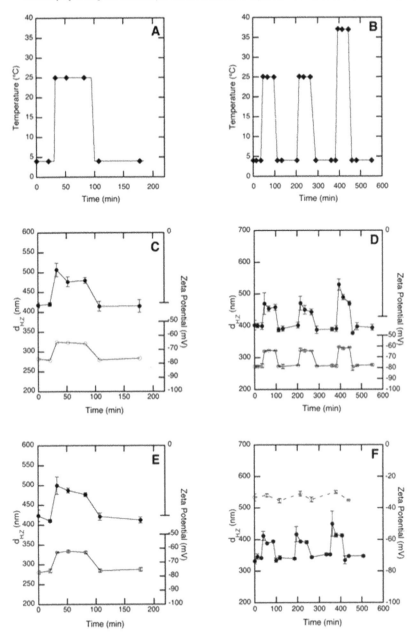

Figure 5 *Effect of the medium temperature on $d_{H,Z}$ (solid circles) and zeta potential (empty circles) of nanogels. (A) Temperature modification for (C) animal and (E) fungal chitosan-TPP/alginate nanogels in water, (B) temperature modification for (D) animal chitosan-TPP/alginate nanogels in water and (F) in PBS. Average of 3 measurements ± SD, representative results of at least two independent experiments*

Na-alg coated nanogels exhibit a negative surface charge between -60 mV and -70mV in H_2O, with no significant variation between the chitosan types (Table 4). This highly negative surface charge ensures good colloidal stability due to sufficient electrostatic repulsion of the nanogel spheres in suspension. When measured in PBS and DMEM, the zeta potentials increased to less negative values between -20 mV and -30 mV. However, during the measurements, the conductivity monitored by the ZetaSizer was indicated as too high for a reliable quantitative measurement. Therefore, these values are indicative and most probably a threshold of the equipment. Nevertheless, the nanogels maintained their negative surface charge and colloidal stability in both PBS and DMEM. The strong influence of the ionic strength on the size and zeta potential of the nanogels was clearly confirmed.

3.3 Temperature Effects in Different Media

Figure 5 illustrates the significant effect of temperature on the size, $d_{H,Z}$ and zeta potential of the nanogels. The $d_{H,Z}$ increases with the temperature and decreases when the temperature is reduced, recovering the initial $d_{H,Z}$ values. The $d_{H,Z}$ measured immediately after reaching the target temperature is higher with a higher standard deviation than subsequent measurements after an equilibration time at the target temperature. The $d_{H,Z}$ then remains stable. Similar behavior as in water is observed in PBS. (Figure 5F). The $d_{H,Z}$ remained stable for several temperature cycles. Animal and fungal chitosan nanogels behave similar upon temperature variations. Moreover, this process is clearly reversible as both the $d_{H,Z}$ and zeta potential recover the initial values when the temperature is again decreased (Figure 5D, F). Even when applying repeated temperature cycles, the initial $d_{H,Z}$ and zeta potential could be reproduced. These observations are important in obtaining the "true" characteristics of nanogels in their application conditions, *in vitro* or *in vivo*.

3.4 Stability Under *in vivo* Simulated Conditions

Three different *in vivo* conditions were simulated by using protein containing media, 1% w/v BSA in DMEM, complete serum supplemented DMEM, and cell culture supernatants. Contrary to chitosan-TPP/alginate nanogels, positively charged chitosan-TPP complexes without Na-alg coating were not stable in cell culture media. Immediate aggregation upon mixing with DMEM only, or supplemented with 10% v/v serum, was observed by DLS, and no meaningful data could be acquired.

The first medium, for which the size distribution was monitored, was DMEM supplemented with 1% w/v BSA to simulate an *in vivo* environment with defined signals (Figure 6A). The nanogels remain stable in the presence of the protein (black line). The peak in the intensity distribution is similar to the nanogel distribution. The second smaller peak is attributed to the BSA, as confirmed by the distribution of the media alone (grey line) and in agreement with the size of the protein.

Nanogels of both animal and fungal chitosan in the second medium, DMEM supplemented with 10% v/v FBS, yielded stable nanogel suspensions as well (Figure 6B). The $d_{H,Z}$ obtained were 168 nm and 156 nm for PdI values of 0.731 and 0.630, for the animal and fungal chitosan nanogels, respectively. These numbers highlight the difficulty of using the $d_{H,Z}$ from DLS when the distribution is composed of more than one size population (peak). In the distribution as intensities, two minor signals at approx. 8 nm and 25 nm appear in the distributions. These small peaks are attributed to the serum proteins. The larger peak of nanogels covering the range of 100 to 1000 nm, corresponds to the size distribution of nanogels in DMEM only (Figure 4E). Animal and fungal-derived nanogels show similar distributions.

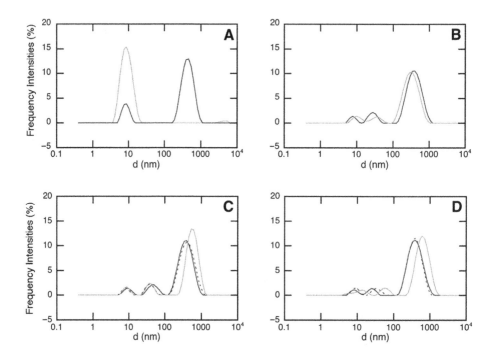

Figure 6 *Stability of chitosan-TPP/alginate nanogels under in vivo relevant conditions, size distributions as intensity. (A) DMEM with 1% v/v BSA, (black line) fungal chitosan nanogels, (grey line) BSA without nanogels, T=37°C. (B) DMEM with 10% w/v FBS, (black line) animal chitosan nanogels, (grey line) fungal chitosan nanogels. (C, D) cell culture supernatants, (C) from HT29 human colon cancer cells, (D) from HCEC human brain derived endothelial cells, (dotted black) freshly mixed 1:1 v/v (measurement at T=25°C), (black solid) after 24 h incubated at 37°C (measurement at T=37°C), (grey solid) after 88 h incubated at 37°C (measurement T=37°C)*

As a third case, nanogels were characterized in cell media supernatants of two different human cell lines, HT-29 and HCEC, employed in cytotoxicological evaluations of these nanogels.[14] The size distribution as intensity remains stable over at least 24 h (Figures 6C and 6D). At 88 h, while no aggregation is observed (gray line), the size distribution is shifted towards larger sizes. Opsonization of the nanogel surface by proteins of the cell supernatant could explain this shift towards higher hydrodynamic diameter values. Although the overall surface charge of the nanogels is strongly negative, it cannot be excluded that chitosan chains penetrate to the surface and interact with negatively charged proteins.

3.5 Mechanical Stability

To illustrate the mechanical stability and deformability of the nanogels, the influence of mechanical stress by filtration on the $d_{H,Z}$ and corresponding size distribution is shown in Figure 7.

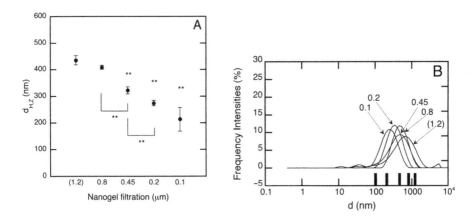

Figure 7 *(A) Hydrodynamic diameter as z-average $d_{H,Z}$, (B) size distributions as intensities for animal chitosan nanogels filtered through 1.2 μm (as standard procedure), and subsequently 0.8 μm, 0.45 μm, 0.2 μm, 0.1 μm. (A) Average of 3 independent samples ± SD. **p<0.01 compared to the diameter of the 1.2 μm fraction. (B) Representative examples of size distributions. The black bars underneath indicate the average pore size of the filter membrane used in sequence of 1.2 to 0.1 μm*

Compared to the initial $d_{H,Z}$ of the nanogels in the suspension prepared in water, the $d_{H,Z}$ decreases significantly for each pore size filtration. However, no cut off is observed. Nanogels with diameters exceeding the pore size of each filter passed through the filters demonstrating the deformability of the nanogels. This observation also signifies that such nanogel suspensions cannot simply be fractionated by filtration. The deformability of the nanogels is important for applications via injection but also for therapies requiring cell uptake. Indeed, the relatively high $d_{H,Z}$ was shown not being a limitation for the nanogels to pass cell membranes.[14]

3.6 Long-term Stability

The durability of nanogels in terms of size and zeta potential was systematically assessed for a period of two months. The nanogels were stored in water and in PBS at either 4°C or room temperature according to the scheme in Figure 2. The results are summarized in Figure 8. During the first week, a slight increase of the $d_{H,Z}$ is observed for the samples kept in water, and for samples in PBS during the first 24 h. After one week, the $d_{H,Z}$ values remained almost unchanged and did not vary within nine weeks. A relatively high deviation was observed, but this is mainly due to a size difference between the two batches used for the study. The standard deviation of the duplicates of each batch was generally small with ±20 nm at the most for either batch.

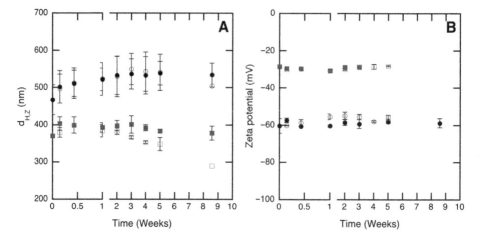

Figure 8 *Stability of chitosan-TPP/alginate nanogels, hydrodynamic diameter (z-average $d_{H,Z}$) and zeta potential as a functionof time. (A) $d_{H,Z}$, (B) zeta potential of BSA-loaded nanogels for different storage media and temperatures. Storage media: (circles) water, (squares) PBS. Nanogels were prepared under sterile conditions and stored in sterilized containers. Temperature of storage: (solid) 4°C, (empty) room temperature. Temperature of measurement: 25°C for all samples. Average of 3 measurements ± SD, representative results of at least two independent experiments. Storage scheme - see Figure 2*

As shown in section 3.2, the $d_{H,Z}$ are significantly lower for nanogels stored in PBS. Size stability was observed during the period of the experiment with the exception of one batch at room temperature at nine weeks for which the z-average decreased. Standard deviations are in this case relatively small, less than ±10 nm for most samples. For nanogels in water, the storage temperature did not have a significant influence. The z-averages measured were similar for samples stored at 4°C or room temperature. The zeta potential remained stable over the period of nine weeks in water and did not vary significantly when stored at 4°C or room temperature.

4 CONCLUSIONS

The application conditions of nanogels frequently differ from those of their preparation and storage. The impact of such differences on the physical properties of nanogels prepared from both animal and fungal chitosan was addressed in this chapter.

Nanogels produced by electrostatic complex formation of chitosan and TPP and subsequently coated with Na-alg are spherical, with hydrodynamic diameters as z-average between 450 and 600 nm in water. Chitosan products derived from both animal and fungal source were able to form nanogels in a comparable manner and with similar nanogel characteristics. The surface charge of nanogels prepared in water is strongly negative with zeta potentials of approx. -65 mV. The hydrodynamic diameter significantly decreases while the zeta potential significantly shifts to less negative values upon transferring the nanogels into PBS or cell culture media. Applying repeated alternating temperature modifications in the range of 4°C to 37°C revealed changes but reversibility of both the hydrodynamic diameter and zeta potential as a function of the temperature. In contrast to uncoated chitosan-TPP complexes, which have a strong aggregation tendency immediately after preparation, the surface modified chitosan-TPP/alginate nanogels are stable in cell culture media and under physiological conditions.

Acknowledgements

This work was funded by the Swiss National Science Foundation and CTI (grant No 404740-117323/1), and supported by Medipol SA (Ecublens, Switzerland).

References

1   M. Rinaudo, *Prog. Polym. Sci.*, 2006, **31**, 603.
2   S. Gautier, *Drug Delivery Technol.*, 2009, **9**, 20.
3   N.S. Rejinold, M. Muthunarayanan, K. Muthuchelian, K.P. Chennazhi, S.V. Nair and R. Jayakumar, *Carbohydr. Polym.*, 2011, **84**, 407.
4   K.A. Howard, S.R. Paludan, M.A. Behlke, F. Besenbacher, B. Deleuran and J. Kjems, *Mol. Ther.*, 2009, **17**, 162.
5   K. Khatri, A.K. Goyal, P.N. Gupta, N. Mishra and S.P. Vyas, *Int. J. Pharm.*, 2008, **354**, 235.
6   K. Debache, C. Kropf, C.A. Schütz, L.J. Harwood, P. Käuper, T. Monney, N. Rossi, C. Laue, K.C. McCullough and A. Hemphill, *Parasite Immunol.*, 2011, **33**, 81.
7   K.I. Draget and C. Taylor, *Food Hydrocolloids*, 2011, **25**, 251.
8   M. Hamidi, A. Azadi and P. Rafiei, *Adv. Drug Delivery Rev.*, 2008, **60**, 1638.
9   R. Mahou, and C. Wandrey, *Macromolecules*, 2010, **43**, 1371.
10 S.K. Motwani, S. Chopra, S. Talegaonkar, K. Kohli, F.J. Ahmad and R.K. Khar, *Eur. J. Pharm. Biopharm.*, 2008, **68**, 513.
11 C. Wandrey and L. Bourdillon, in *Analytical Ultracentrifugation - Techniques and Methods*, ed. D.J. Scott, S.E. Harding and A.J. Rowe, Royal Society of Chemistry, Cambridge, 2005, ch. 9, p. 162.
12 G.A. Morris, J. Castile, A. Smith, G.G. Adams and S.E. Harding, *Carbohydr. Polym.*, 2011, **84**, 1430.
13 M.L. Tsai, R.H. Chen, S.W. Bai and W.Y. Chen, *Carbohydr. Polym.*, 2011, **84**, 756.
14 C.A. Schütz, L. Juillerat-Jeanneret, P. Käuper and C. Wandrey, *Biomacromolecules*, 2011, **12**, 4153.
15 C.A. Schütz, F. Schmitt, L. Juillerat-Jeanneret and C. Wandrey, *Chimia*, 2009, **63**, 220.
16 E. Fernandez-Megia, R. Novoa-Carballal, E. Quinoa and R. Riguera, *Carbohydr. Polym.*, 2005, **61**, 155.
17 M. Lavertu, Z. Xia, A.N. Serreqi, M. Berrada, A. Rodrigues, D. Wang, M.D. Buschmann and A. Gupta, *J. Pharm. Biomed. Anal.*, 2003, **32**, 1149.
18 K.M. Vårum, M.W. Anthonsen, H. Grasdalen and O. Smidsrød, *Carbohydr. Res.*, 1991, **211**, 17.
19 M.L. Huggins, *J. Am. Chem. Soc.*, 1942, **64**, 2716.
20 P. Schuck, version 11.8.
21 P. Käuper and C. Laue, Patent number WO2007031812, 2007.
22 G.M. Pavlov and S.G. Selyunin, *Vysokomol. Soedin., Ser. A*, 1986, **28**, 1727.

23   J.M. Creeth and C.G. Knight, *Biochim. Biophys. Acta, Biophys. Incl. Photosynth.*, 1965, **102**, 549.
24   M. Wales and K.E.V. Holde, *J. Polym. Sci.*, 1954, **14**, 81.
25   G.A. Morris, J. Castile, A. Smith, G.G. Adams and S.E. Harding, *Carbohydr. Polym.*, 2009, **76**, 616.
26   S.E. Harding, in *Analytical Ultracentrifugation in Biochemistry and Polymer Science*, ed. S.E. Harding, A.J. Rowe and J.C. Horton, Royal Society of Chemistry, Cambridge, 1992, ch. 27, p. 495.

10

CELLULOSE CRYSTALLINITY: PERSPECTIVES FROM SPECTROSCOPY AND DIFFRACTION

M.C. Jarvis

School of Chemistry, University of Glasgow, Glasgow G12 8QQ, Scotland, United Kingdom
mikej@chem.gla.ac.uk

1 INTRODUCTION

The idea of crystallinity, when applied to cellulose, is relevant to the mechanical strength of living plants and of materials like wood and cotton that are made from plants, and how these properties can change through degradation and other effects[1]. It is also highly relevant to the conversion of lignocellulosic materials to liquid biofuels[2] and nanostructured fibres for the manufacture of new, sustainable composite materials[3]. This Chapter will demonstrate that the nature of crystallinity in cellulose is not coherently enough understood to meet the requirements that flow from these diverse areas of relevance, and that new approaches are needed.

A perfect crystal has every atom placed on an infinite lattice constructed from a unit cell that is normally of nanometre dimensions. Each atom deviates from its position in the lattice only through thermal fluctuations. Cellulose, however, does not form perfect crystals. It forms fibres called microfibrils. Cellulose microfibrils are long, comprising thousands to millions of unit cells, but their width is only a few unit cells[4]. Cellulose microfibrils also contain various kinds of disorder that are only partially understood[5].

In a broad sense, the crystallinity of any cellulosic material is the extent to which its structure corresponds to that of a perfect crystalline lattice; that is, crystallinity decreases with the extent of disorder within the material. However different kinds of disorder are detected by different experimental methods[5], and different kinds of disorder are relevant to different properties and technological uses of cellulose[6-10]. Some kinds of disorder are major, others are minor.

Against this background, the quantitative concept of cellulose crystallinity has generally been linked to the mass fraction of a cellulosic material that can be described as conforming approximately to the crystalline lattice, if one ignores forms of disorder that are minor or are not detected by whatever method is being used to characterise the structure[11]. It follows that cellulose crystallinity is a method-dependent concept, and that whatever measurement method is used, a judgement must be made about what forms of disorder are 'minor' enough to be safely ignored. Also, the crystallinity concept is best expressed in a variety of ways appropriate to different end uses of the cellulose, because each end use is affected by different forms of disorder in the structure[3,12].

This framework will be used to explore the capacity of various spectroscopic and scattering methods to detect disorder of different forms; and then to ask the question which of these forms of disorder (non-crystallinity) are relevant to major applications of cellulose. First, however it is necessary to summarise current ideas on the structure of cellulose microfibrils and the extent of disorder within them.

2   ORDER AND DISORDER IN THE STRUCTURE OF CELLULOSE MICROFIBRILS FROM HIGHER PLANTS

Until quite recently there was a widespread view that in most plant materials cellulose microfibrils were of constant size, containing 36 chains in a diamond-shaped configuration with just over half of these chains exposed at the surface (Figure 1). A multiple of six for the chain number was preferred because of the hexameric structure of the terminal complex from which each nascent microfibril emerges[13]. The 'crystalline' fraction was thought to contain variable proportions of the two allomorphs cellulose Iα and Iβ[4,14] distinguished mainly by the longitudinal stagger between sheets of glucan chains (Figure 1) and accompanied by a considerable amount of disordered cellulose[15]. The NMR community equated the disordered chains with those at the microfibril surface[16,17] whereas in the biofuels research community it was widely held that disordered and crystalline domains alternated along the microfibril[12], a view originating in earlier crystallographic publications[18].

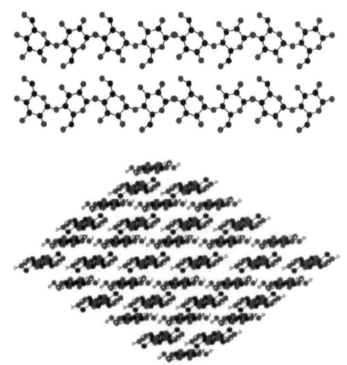

Figure 1 *Commonly accepted ideas of the structure of cellulose microfibrils. Top: two cellulose chains hydrogen-bonded together as in the cellulose Iβ crystalline lattice. Bottom: 36-chain diamond-shaped microfibril in cross section*

A recent publication on spruce cellulose[5] questioned some of these ideas, confirming the widely accepted microfibril diameter of 3.0 nm and pointing out that this diameter left room insufficient room for 36 chains. A group of 24-chain models were proposed, with the additional possibility of microfibrils containing 18 cellulose chains plus a small amount of tightly bound glucomannan. 'Square' configurations with the hydrophobic [200] crystal face exposed fitted some of the data better than the 'diamond' configuration found in algal celluloses. Conformationally disordered cellulose was located predominantly, but probably not wholly, at the surface. The 'crystalline' core was insufficiently well-ordered for the distinction between the Iα and Iβ forms to be meaningful[5].

It is uncertain to what extent cellulose microfibrils from other higher plant materials resemble spruce microfibrils. Primary-wall cellulose is known to be more disordered than wood cellulose and to have somewhat different unit-cell dimensions[19]. In flax, cotton and tension wood cellulose, increased microfibril diameter is accompanied by a smaller fraction of conformationally disordered chains and a preponderance of the Iβ allomorph[15,20-23].

Higher plants have distinct sets of cellulose synthase gene products synthesising primary-wall and secondary-wall cellulose[13]. It is unclear to what extent differences in microfibril structure result from differences in the synthetic enzyme systems or from differences in the self-assembly of microfibrils in varied cell-wall environments, but coalescence during self-assembly seems the most likely origin of the large microfibrils of the textile celluloses and tension wood.

In conifer wood, cellulose microfibrils aggregate laterally into bundles that can be penetrated only partially by water[5,24-27]. It is not clear to what extent aggregation also occurs in other cell-wall types, but it has been observed in the hardwood poplar[28]. Less aggregation seems to occur in tension wood[28] perhaps because the microfibrils coalesce instead during their formation.

Following this very brief summary of current ideas on microfibril structure, the methods for deriving structural information will now be described, with emphasis on distinguishing order from disorder.

3 METHODS FOR ASSESSING CRYSTALLINITY AND DISORDER

3.1 X-ray and neutron diffraction

Cellulose gives clear X-ray diffraction patterns[4] and it seems logical that this method should be used to quantify the 'crystalline' cellulose component in cellulosic materials. There is a long history of its use for this purpose and standardised methods are well established[11]. Nevertheless practical problems remain and it is by no means clear what properties of the material distinguish 'crystalline' from 'non-crystalline'. The background to the x-ray methods for quantifying crystallinity together with many of the difficulties have been reviewed[11]. Briefly, it is most common to use powder diffraction methods and to assume that sharp reflections correspond to crystalline cellulose, whereas diffuse scattered intensity comes from non-cellulosic components together with whatever non-crystalline cellulose is present.

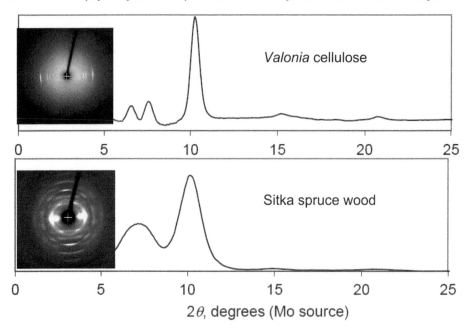

Figure 2 *Wide-angle X-ray scattering patterns from highly crystalline* Valonia *cellulose (top) and Spruce wood (bottom) The scattering pattern from spruce wood shows much more broadening in the radial direction. The radial broadening is due to both disorder and the small lateral dimensions of the microfibrils*

A conceptual distinction must be made between the crystallinity of a material like wood and the crystallinity of the cellulose within the material. The crystallinity of wood is the mass of crystalline cellulose as a fraction of the total mass, about half of which comprises non-cellulosic, non-crystalline components[11]. This is what is measured by diffraction methods (Figure 2). It is therefore meaningful to speak of the crystallinity of the cellulose itself only when pure cellulose is being measured or if an appropriate correction for the non-cellulosic mass fraction can be made.

There are also some practical problems in separating the sharp and diffuse components and a difficulty in finding an appropriate standard for the quantification of the non-crystalline component[18,29]. The use of Rietveld refinement may facilitate a direct solution to this problem[30] but it has emerged that bound water makes a surprisingly large contribution to the diffuse scattering[31]. Because effects of microfibril orientation are averaged out in powder diffraction the principal structural feature that determines whether or not scattered peaks are sharp is the regularity of lateral chain packing, and this is what 'crystallinity', as measured in this way[11] really signifies.

Diffraction patterns contain much more information than is used in the standard procedures for crystallinity measurement, especially when the cellulose is sufficiently well-oriented to justify the analysis of fibre rather than powder diffraction methods[18]. The crystal structures of the two native allomorphs, cellulose Iα and Iβ, were solved by a combination of X-ray and neutron diffraction on well-oriented and highly crystalline algal and tunicate celluloses[32,33]. The neutron diffraction experiments allowed protons to be located within each structure, with sufficient precision to show that the internal hydrogen-bonding schemes are somewhat disordered. That is, two alternative hydrogen-bond networks were fractionally occupied in each allomorph[32,33].

There is also a long history of the assessment of disorder from radial broadening of the X-ray reflections[34]. Radial broadening results directly from small lattice dimensions (the Scherrer effect) as well as from disorder[18], and it is particularly difficult to separate these contributions when, as is common, there is enough disorder to eliminate most higher-order reflections[35]. A partial solution to this problem was recently presented and allowed 'paracrystalline' disorder and at least one other form of disorder in lateral chain packing to be quantified even in the presence of significant Scherrer broadening[5]. Longitudinal disorder can be estimated in a similar way[35] but is much smaller in magnitude than lateral disorder. Wood cellulose also shows signs of diagonal disorder resulting from irregularity in the longitudinal inter-sheet stagger that differentiates the Iα and Iβ allomorphs[5].

The diffraction methods in common use include the estimation of the mass fraction of crystalline cellulose by semi-empirical powder methods and the estimation of microfibril diameter through the Scherrer Equation. The other approaches mentioned above have not been widely exploited.

3.2 ^{13}C NMR spectroscopy

The first evidence that native cellulose comprised two allomorphs came from solid-state ^{13}C NMR spectroscopy[36]. It has also been clear from the earliest applications of this technique that many chains in cellulose from higher plants differ from the conformation found in the most crystalline algal and tunicate celluloses[37]. The conformational differences are evident in the C-4 and C-6 signals[17] (Figure 3) Signals assigned to these 'non-crystalline' forms become more abundant the smaller the microfibril diameter as assessed by EM or diffraction methods[38,39]. This observation, and the evidence from spin-diffusion and ^{13}C relaxation experiments[5,39-41], are all consistent with a surface location for most of the 'non-crystalline' chains rather than a model in which these occur in blocks occupying the whole width of the microfibril and alternating longitudinally with crystalline regions. Some of the 'non-crystalline' chains appear to be on surfaces that are exposed to water and organic solvents, while others are on surfaces buried within aggregates of microfibrils and are inaccessible to solvents[26].

In principle signal intensities obtained by the routine CP-MAS (cross-polarisation, magic-angle spinning) experiment are not quantitative but in practice for cellulose – not for soluble polysaccharides or lignin – quantitation is quite accurate[42]. There is some difficulty in separating xylan, xyloglucan and glucomannan signals from the signals assigned to disordered cellulose, and relaxation experiments[38] and deconvolution procedures[26] have been devised to circumvent this problem.

The correlation between the fractional abundance of conformationally distinct chains, apparently at microfibril surfaces, with microfibril diameter as estimated crystallographically though the Scherrer equation[38] lends credibility to the idea that chains located at surfaces are in some way different from chains within[17]. These parameters do also correlate in some circumstances with crystallinity as measured by diffraction methods[18], but it would be unwise to assume that the chains classed as non-crystalline by NMR are also in the non-crystalline fraction as assessed by X-ray diffraction. When the NMR method depends on chain conformation and the diffraction method depends on chain packing, they need not 'see' the same chains as being different from crystalline cellulose.

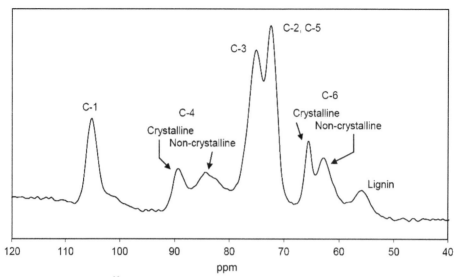

Figure 3 *Solid-state ^{13}C NMR spectrum of spruce wood. The C-4 and C-6 signals from crystalline and disordered cellulose are distinct from one another due to differences in time-averaged conformation between the crystalline and disordered forms*

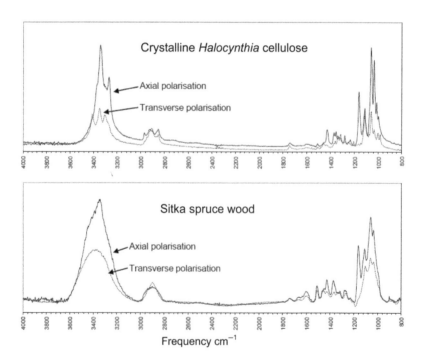

Figure 4 *Polarised FTIR spectra from highly crystalline* Halocynthia *(tunicate) cellulose and spruce wood. The* Halocynthia *spectrum is much better resolved than the spruce wood spectrum, due to greater uniformity of structure and of hydrogen bonding*

3.3 FTIR spectroscopy

The hydroxyl stretching region of the FTIR spectrum carries detailed information on hydrogen-bonding networks in cellulose[43]. The challenge is to interpret this information. Hydroxyl stretching modes are not complicated so much by coupling as lower-frequency modes are, but complete assignments are not yet available even for the six distinct hydroxyl groups in either of the hydrogen-bond network in either one of the two native cellulose allomorphs[39,43]. Nevertheless the distinctive patterns of OH stretching bands associated with both allomorphs are known[43,44] and OH stretching intensity that does not conform to these patterns may be assigned to 'disordered' cellulose, differing from the accepted forms of crystalline cellulose with respect to its hydrogen bonding[39,45].

Three developments of this technique permit further useful information to be obtained. First, polarisation allows the direction in which each hydroxyl group points to be identified[39,45] (Figure 4). In most studies only longitudinal and transverse polarisation have been distinguished, but a quantitative treatment is available[21]. Secondly, substitution with D_2O allows accessible chains to be distinguished from chains within the crystal lattice, since the hydroxyl bands are moved to lower frequency by a factor of 1.34, into an otherwise empty spectral region[39,44]. This experiment clearly demonstrates that surface chains differ in hydrogen bonding from crystalline cellulose, as would be expected from the different C-6 conformation shown by ^{13}C nuclear magnetic resonance (NMR)[5,39]. Some chains with similar hydrogen-bonding patterns remain inaccessible to D_2O, consistent with water-impermeable lateral aggregation of cellulose microfibrils as shown by NMR[5]. Thirdly, by combining polarisation with oscillating mechanical stress in a two-dimensional experiment it is possible to demonstrate which chains are mechanically linked to which[46]. Tight mechanical association between crystalline cellulose and disordered cellulose chains accessible to D_2O was consistent with side-by-side bonding of these chains and was not readily reconciled with models in which crystalline and disordered (*sensu* D_2O-accessible) chains were arranged in series along each microfibril[47].

It is clear that all the techniques described above measure non-crystalline cellulose of one kind or another. But no two techniques measure the same deviations from the structure of perfectly crystalline cellulose. To interpret the results obtained by any technique it is first necessary to ask what kind of disorder is important, in the context of the use to which the cellulose will be put.

4   WHAT KINDS OF DISORDER MATTER?

4.1   Cellulose crystallinity in living plants.

In growing plant cells, mechanical strength and rigidity come from the primary cell wall inflated by the turgor (osmotic) pressure within the cell[48]. The primary cell wall is thin, lamellate and composed of a partially oriented net of cellulose microfibrils embedded in a hydrated matrix of other polysaccharides[49]. The statistical orientation of the restraining microfibrils controls the direction of growth and hence the development of plant form[50]. Primary-wall microfibrils are about 3 nm or less in diameter and contain considerable disorder as detected by almost all measurement techniques[19,39,51-53]. They interact non-covalently with a highly disordered matrix[49]. In contrast wood, and the woody cell walls of plant stems and veins, do not grow and have a much thicker secondary cell wall deposited inside the primary wall[54]. Secondary-wall cellulose microfibrils are embedded in a lignified matrix and are oriented in a helix whose pitch determines the longitudinal stiffness of the plant tissue and is under developmental control[55] depending on whether rigidity or flexibility is a more appropriate reaction to forces like wind and gravity[56]. The microfibrils of wood cellulose are about 3 nm in diameter[5], like those of primary cell walls, but their crystallinity as measured by most techniques is a little greater[34].

There are a number of specialised plant tissues in which the cellulose microfibrils are of larger diameter, typically 5 nm, and are more crystalline however crystallinity is measured. These tissues include bast fibres such as flax (linen), ramie and hemp[22,39]; the tension wood that keeps branches at the right angle in hardwood trees[10,23]; and seed hairs such as cotton[20,22,26]. These are anatomically unrelated but all have cell walls in which the fraction of non-cellulosic material is much less than in primary cell walls or in wood, so that cellulose is the dominant constituent. It seems that in these circumstances cellulose microfibrils fuse into larger units as they are formed[39].

In these plant materials the cellulose is loaded principally in tension. It may be hypothesised that the increased microfibril diameter and crystallinity are adaptations to increase the tensile performance, and there are indications that tensile stiffness and strength are indeed higher in these forms of cellulose than in wood cellulose[57,58].

Much more crystalline microfibrils, with much greater lateral dimensions up to 20 nm, are found in some algae and in tunicates, that only animals that make cellulose[32,33]. Bacterial cellulose is also more crystalline than cellulose in higher plants[14,22]. NMR and FTIR spectroscopy show that in the small microfibrils from higher plants the polymer chains exposed at the cellulose surface differ conformationally, and have less hydrogen bonding directed along each chain, than in the more crystalline interior[16,39]. In large algal microfibrils this is apparently not the case[59], perhaps because different crystallographic planes are exposed or because the wider surface planes give less opportunity for lateral chain packing to expand.

The tensile modulus of crystalline cellulose is computed as about 140 GPa[60] making cellulose a high-stiffness material, at least in relation to its weight. Modelling studies show that hydrogen bonding makes a remarkably large contribution to the tensile modulus, more than would be expected from the longitudinal stiffness of the hydrogen bonds themselves on a simple additive model[61,62]. The nature of this phenomenon is not clear, but it is consistent with the hypothesis that microfibrils of larger diameter in the textile celluloses have superior tensile performance. Surface chains, in addition to having hydrogen bonds directed outward towards the matrix rather than along the microfibril, are partially exposed to hydrogen-bond cleavage through insertion of water[16].

Mutants defective in cellulose synthesis show impaired growth and reduced cellulose content in the walls of affected cells[13], but in most cases the cellulose is structurally normal. In a small number of exceptions, reduced crystallinity has been claimed on the basis of evidence from X-ray diffraction[7,8,63,64] and NMR[63]. Because both disordered chain packing and small lateral dimensions can contribute to radial broadening in diffraction patterns it would be of interest to examine these mutants by additional methods.

4.2 Cellulose in structural composite materials.

There is rapidly growing interest in the use of nanocellulose products as fibre reinforcements for sustainable composite materials[3]. Here, unlike the environment of living plants where cellulose evolved, there is no water and hydrogen-bond hydrolysis is not an issue: however the orientation of the hydrogen bonding in the surface chains may still affect the mechanical performance of the fibres (and their interactions with what is not necessarily a hydrophilic matrix). The performance of composites containing whiskers of tunicin, which has particularly wide, highly ordered microfibrils, was consistent[65] with the calculated tensile modulus of approximately 140 GPa for fully crystalline cellulose[62]. These calculations suggested that inter-residue hydrogen bonds made a considerable contribution to the longitudinal modulus of cellulose Iβ, which might therefore be reduced in the surface chains of wood cellulose[60]. Here and within living plants, the key aspects of cellulose crystallinity are the fraction of surface chains where these differ in conformation and in orientation of their hydrogen bonds – details that are best detected by NMR and polarised FTIR respectively.

The manufacture of cellulose whiskers from sustainable plant sources of cellulose depends on another form of disorder. Acid hydrolysis of cellulose microfibrils, particularly when preceded by sulphation with sulphuric acid, cleaves the microfibrils at regular intervals into sections a few hundred nanometres in length[66]. It would seem that the cleavage points are marked by pre-existing domains that differ in structure from the rest of the microfibril and thus class as disordered[66], but the nature of the disorder is unclear. The known length of individual cellulose molecules in higher plants would lead to gaps between chain ends at approximately the right spacing if these were regularly staggered along the microfibril[67], but suitable methods for detecting such gaps have not been developed for application to cellulose from plant sources. The relevant form of disorder in this case is therefore uncharacterised.

A widely adopted model for cellulose microfibrils locates the non-crystalline material in blocks alternating with crystalline regions along each microfibril. This model was originally based on the limited crystallographic coherence in the direction of the microfibril axis[66] and was supported by crystallographic evidence that tensile stress was distributed equally between crystalline and non-crystalline domains[68]. However there is extensive NMR and FTIR data favouring a surface location for most of the disordered material detectable by these methods, and other explanations have now been offered for the short crystallographic coherence length[5]. The evidence available therefore no longer supports the idea that most of the disordered cellulose occurs in large blocks alternating longitudinally with the crystalline cellulose. However there is no evidence against the presence of small disordered domains within microfibrils, accounting for a minor fraction of the disordered material and associated for example with chain ends.

4.3 Cellulose degradation in biofuel production.

Technologies now being evaluated for conversion of lignocellulose substrates to liquid biofuels involve a chemical pre-treatment to dissociate lignin from cellulose, followed by saccharification (depolymerisation) of cellulose and other polysaccharides and fermentation to biofuel[2]. The saccharification step may be carried out by added cellulolytic enzymes, or these enzymes may be secreted by the organism used for fermentation. In either case, saccharification is a major rate-limiting step[69]. The recalcitrance of the cellulosic material is a key determinant of the suitability of different biomass materials for biofuel production and much of the technological effort is devoted to pre-treatments that alleviate this recalcitrance[69].

Cellulose, as a more or less crystalline solid, is not evidently amenable to degradation by soluble enzymes. Saccharification of cellulose is therefore normally a multistep process. The first enzymatic step is normally attack on one of the exposed faces of a cellulose microfibril[69]. The initial attacking enzyme may be an endo-hydrolase, and one group of such endo-hydrolases contain binding domains that lock onto the relatively hydrophobic [200] crystal plane of cellulose[70]. Another group of hydrolases have a binding site that wraps around a single cellulose chain and therefore must first detach a short length of this chain from the microfibril surface[69]. A recently discovered group of enzymes including fungal GH61 glycoside hydrolases[71] and bacterial CBM33 proteins[72] cleave cellulose chains by an oxidative mechanism. One group of these enzymes, secreted extensively by wood-rotting fungi, have an active site located in the centre of a large flat face that apparently binds to one of the exposed cellulose crystal planes[71].

The accessibility of cellulose microfibrils to degradation may be expected, therefore, to depend on the enzyme system used for the initial hydrolytic or oxidative attack on the solid cellulose surface. It is of importance to know which surfaces are exposed[73] and how the chains on these surfaces differ, in packing, conformation and hydrogen bonding, from chains within the crystalline lattice[5]. Disordered domains in which some chains are only loosely attached to the surface would be particularly relevant[74,75]. For the rather disordered celluloses of lignocellulosic biomass this information is difficult to access.

5 CONCLUSIONS

The disorder incorporated into the structure of any cellulose microfibril influences the performance of the cellulose, whether in the living plant, in sustainable composites or during conversion to biofuel. In each of these cases the kind of disorder that matters is different and needs to be quantified by different methods. It is not sensible, at present, to have a single definition of crystallinity nor to measure it by a single method.

When the question is the kind of cellulose that is made by different plant tissues, variation in microfibril diameter and hence in surface:volume ratio is the main issue. Scherrer broadening in X-ray diffraction may be used as a measure of diameter or the surface: volume ratio can be accessed by ^{13}C NMR. The nature of the reduced crystallinity in certain mutants with impaired cellulose synthesis needs to be more clearly defined before the most appropriate technique for quantification can be identified.

References

1   I. Burgert, and P. Fratzl, *Integrative Comparative Biol.*, 2009, 49, 69.
2   A. Carroll, and C. Somerville, *Ann. Rev. Plant Biol.*, 2009, 60, 165.
3   S.J. Eichhorn, A. Dufresne, M. Aranguren, N.E. Marcovich, J.R. Capadona, S.J. Rowan, C. Weder, W. Thielemans, M. Roman, S. Renneckar, W. Gindl, S. Veigel, J. Keckes, H. Yano, K. Abe, M. Nogi, A.N. Nakagaito, A.Mangalam, J. Simonsen, A.S. Benight, A. Bismarck, L. Berglund, and T. Peijs, *J. Mater. Sci.*, 2010, 45, 1.
4   Y. Nishiyama, *J. Wood Sci.*, 2009, 55, 241.
5   A.N. Fernandes, L.H. Thomas, C.M. Altaner, P. Callow, V.T. Forsyth, D.C. Apperley, C.J. Kennedy and M.C., *Proc. Nat. Acad. Sci., USA*, 2011, 108, E1195.
6   S.P.S. Chundawat, G. Bellesia, N. Uppugundla, L.D. Sousa, D.H. Gao, A.M. Cheh, U.P. Agarwal, C.M. Bianchetti, G.N. Phillips, P. Langan, V. Balan, S. Gnanakaran and B.E. Dale, *J. Am. Chem. Soc.*, 2011, 133, 11163.
7   M. Fujita, R. Himmelspach, C.H. Hocart, R.E. Williamson, S.D. Mansfield, and G.O. Wasteneys, *Plant J.*, 2011, 66, 915.
8   D. Harris, J. Stork, J. and S. Debolt, *Global Change Biol. Bioenergy*, 2009, 1, 51.
9   Y.C. Hsieh, H. Yano, M. Nogi and S.J. Eichhorn, *Cellulose*, 2008, 15, 507.
10 M. Wada, T. Okano, J. Sugiyama and F. Horii, *Cellulose*, 1995, 2, 223.
11 A. Thygesen, J. Oddershede, H. Lilholt, A.B. Thomsen, and K. Stahl, *Cellulose*, 2005, 12, 563.
12 D. Harris and S. Debolt, *Plos One*, 2008, 3, E2897.
13 C. Somerville, *Ann. Rev. Cell Development. Biol.*, 2006, 22, 53.

14 R.H. Newman, *Holzforschung,* 1999, **53,** 335.
15 M. Jarvis, *Nature,* 2003, **426,** 611
16 R.H. Newman and T.C. Davidson, *Cellulose,* 2004, **11,** 23.
17 R.J. Vietor, R. H. Newman, M.A. Ha, D.C. Apperley and M.C. Jarvis, *Plant J.,* 2002, **30,** 721.
18 S. Andersson, R. Serimaa, T. Paakkari, P. Saranpaa, and E. Pesonen, *J. Wood Sci.,* 2003, **49,** 531.
    19 E. Dinand, H. Chanzy, H. and M.R. Vignon, *Cellulose,* 1996, 3, 183.
20 K. Leppanen, S. Andersson, M. Torkkeli, M. Knaapila, N. Kotelnikova and R. Serimaa, *Cellulose,* 2009, **16,** 999.
21 A. Sturcova, I. His, T.J. Wess, G. Cameron, and M.C. Jarvis, *Biomacromol.,* 2003, **4,** 1589.
22 M. Wada, T. Okano and J. Sugiyama, *J. Wood Sci.,* 2001, **47,** 124.
23 R. Washusen and R. Evans, *IAWA Journal,* 2001, **22,** 235.
24 J. Fahlen, and L. Salmen, *Biomacromol.,* 2005, 6, 433.
25 T. Liitia, S.L. Maunu, B. Hortling, T. Tamminen, O. Pekkala and A. Varhimo, *Cellulose,* 2003, **10,** 307.
26 K. Wickholm, P.T. Larsson and T. Iversen, *Carbohyd. Res.,* 1998, **312,** 123.
27 P. Xu, L.A. Donaldson, Z.R. Gergely and L.A. Staehelin, L. A., *Wood Sci. Technol.,* 2007, **41,** 101.
28 L. Donaldson, *Wood Sci. Tech.,* 2007, **41,** 443.
29 S. Park, J.O. Baker, M.E. Himmel, P.A. Parilla and D.K. Johnson, *Biotech. for Biofuels,* 2010, **3,** 10.
30 C. Driemeier and G.A. Calligaris, *J. Appl. Crystallog.,* 2011, **44,** 184.
31 S. Zabler, O. Paris, I. Burgert and P. Fratzl, *J. Struct. Biol.,* 2010, **171,** 133.
32 Y. Nishiyama, P. Langan and H. Chanzy, *J. Am. Chem. Soc.,* 2002, **124,** 9074.
33 Y. Nishiyama, J. Sugiyama, H. Chanzy and P. Langan, *J. Am. Chem. Soc.,* 2003, **125,** 14300.
34 H.F. Jakob, D. Fengel, S.E. Tschegg and P. Fratzl, *Macromolecules,* 1995, **28,** 8782.
35 A.M. Hindeleh and R. Hosemann, *J. Mater. Sci.,* 1991, **26,** 5127.
36 R.H. Atalla, and D.L. Vanderhart, *Solid State Nucl. Magnet. Resonan.,* 1999, **15,** 1.
37 F. Horii, A. Hirai, and R. Kitamaru, *J. Carbohyd. Chem.,* 1984, **3,** 641.
38 R.H. Newman, *Solid State Nucl. Magnet. Resonan.,* 1999, **15,** 21.
39 A. Sturcova, I. His, D.C. Apperley, J. Sugiyama and M.C. Jarvis, *Biomacromol.,* 2004, **5,** 1333.
40 C. Altaner, D.C. Apperley, and M.C. Jarvis, *Holzforschung,* 2006, **60,** 665.
41 M. Foston, R. Katahira, E. Gjersing, M.F. Davis and A. J. Ragauskas, *J. Agric. Food Chem.,* 2012, **60,** 1419.
42 R.H. Newman and J.A. Hemmingson, *Cellulose,* 1995, **2,** 95.
43 Y. Marechal and H. Chanzy, *J. Mol. Struct.,* 2000, **523,** 183.
44 Y. Horikawa, B. Clair, B. and J. Sugiyama, *Cellulose,* 2009, **16,** 1.
45 Y. Hishikawa, E. Togawa, and T. Kondo, *Cellulose,* 2010, **17,** 539.
46 M. Akerholm, B. Hinterstoisser and L. Salmen, *Carbohyd. Res.,* 2004, **339,** 569.
47 L. Salmen and E. Bergstrom, *Cellulose,* 2009, **16,** 975.
48 D.J. Cosgrove, *Nature Rev. Mol. Cell Biol.,* 2005, **6,** 850.
49 M.C. Jarvis, *Food Hydrocoll.,* 2011, **25,** 257.
50 T.I. Baskin, *Ann. Rev. Cell Development. Biol.* 2005, **21,** 203.
51 S.Y. Ding and M.E. Himmel, *J. Agric. Food Chem.,* 2006, **54,** 597.
52 L.D. Melton, I. Kurtovic and B.G. Smith, *Mitteilungen der Bundesforschungsanstalt für Forst- und Holzwirtschaft, Hamburg,* 2007, 21.

53 H. Niimura, T. Yokoyama, S. Kimura, Y. Matsumoto, and S. Kuga, *Cellulose,* 2010, **17,** 13.
54 P. Fratzl and R. Weinkamer, *Prog. Materials Sci.,* 2007, **52,** 1263.
55 E.J. Mellerowicz and B. Sundberg, *Curr. Opin. Plant Biol.,* 2008, **11,** 293.
56 C.M. Altaner, and M.C. Jarvis, *J. Theor. Biol.,* 2008, **53,** 434.
57 A. Ishikawa, T. Okano and J. Sugiyama, *Polymer,* 1997, **38,** 463.
58 I. Diddens, B. Murphy, M. Krisch, and M. Muller, M. *Macromolecules,* 2008, **41,** 9755.
59 A.A. Baker, W. Helbert, J. Sugiyama and M.J. Miles, *J. Struct. Biol.,* 1997, **119,** 129.
60 L. Salmen, *Comptes Rendus Biologies,* 2004, 327, **873.**
61 S.J. Eichhorn and G.R. Davies, *Cellulose,* 2006, **13,** 291.
62 M.S. Cintron, G.P. Johnson, and A.D. French, *Cellulose,* 2011, **18,** 505.
63 D.M. Harris, K. Corbin, T. Wang, R. Gutierrez, A.L. Bertolo, C. Petti, D.-M. Smilgies, J. Manuel Estevez, D. Bonetta, B.R. Urbanowicz, D.W. Ehrhardt, C.R. Somerville, J.K.C. Rose, M. Hong, M. and S. Debolt, *Proc. Nat. Acad. Sci., USA,* 2012, **109,** 4098.
64 C. Sanchez-Rodriguez, S. Bauer, K. Hematy, F. Saxe, A.B. Ibanez, V. Vodermaier, C. Konlechner, A. Sampathkumar, M. Rueggeberg, E. Aichinger, L. Neumetzler, I. Burgert, C. Somerville, M.-T. Hauser and S. Persson, *Plant Cell,* 2012, **24,** 589.
65 A. Sturcova, G.R. Davies and S.J. Eichhorn, *Biomacromol.,* 2005, **6,** 1055.
66 S. Elazzouzi-Hafraoui, Y. Nishiyama, J.L. Putaux, L. Heux, F. Dubreuil and C. Rochas, *Biomacromol.,* 2008, **9,** 57.
67 R. Berggren, F. Berthold, E. Sjoholm and M. Lindstrom, *J. Appl. Polym. Sci.,* 2003, **88,** 1170.
68 K. Kong, M.A. Wilding, R.N. Ibbett and S.J. Eichhorn, *Faraday Discuss.,* 2008, **139,** 283.
69 D.B. Wilson, *Curr. Opin. Biotech.,* 2009, **20,** 295.
70 D.J. Dagel, Y.S. Liu, L.L. Zhong, Y.H. Luo, M.E. Himmel, Q. Xu, Y.N. Zeng, S.Y. Ding and S. Smith, *J. Phys. Chem. B,* 2011, **115,** 635.
71 Z. Forsberg, G. Vaaje-Kolstad, B. Westereng, A.C. Bunaes, Y. Stenstrom, A. Mackenzie, M. Sorlie, S.J. Horn and V.G.H. Eijsink, *Prot. Sci.,* 2011, **20,** 1479.
72 R.J. Quinlan, M.D. Sweeney, L. Lo Leggio, H. Otten, J.-C.N. Poulsen, K.S. Johansen, K.B.R.M. Krogh, C. Jorgensen, M. Tovborg, A. Anthonsen, T. Tryfona, C.P. Walter, P. Dupree, F. Xu, G.J. Davies and P.H. Walton, *Proc. Nat. Acad. Sci., USA,* 2011, **108,** 15079.
73 Q. Xu, M.P. Tucker, P. Arenkiel, X. Ai, G. Rumbles, J. Sugiyama, M.E. Himmel and S.Y. Ding, *Cellulose,* 2009, **16,** 19.
74 C.M. Payne, M.E. Himmel, M.F. Crowley and G.T. Beckham, *J. Phys. Chem. Lett.,* 2011, **2,** 1546.
75 L.G. Thygesen, B.J. Hidayat, K. Johansen and C. Felby, *J. Ind. Microbiol. Biotech.,* 2011, **38,** 975.

11
THE WATER VAPOUR SORPTION PROPERTIES OF CELLULOSE

C.A.S. Hill[1*] and Y. Xie[2]

[1]Forest Products Research Institute, Joint Research Institute for Civil and Environmental Engineering, Edinburgh Napier University, Merchiston Campus, Colinton Road, Edinburgh, EH10 5DT, UK
[2]Key Laboratory of Bio-based Material Science and Technology, Northeast Forestry University, 26 Hexing Road, Harbin, 150040, People's Republic of China
* c.hill@napier.ac.uk

1 INTRODUCTION

The water sorption behaviour of cellulose is discussed in this chapter, both from the point of view of the equilibrium conditions that constitute the sorption isotherm and also the sorption kinetic behaviour. A model for explaining sorption hysteresis is presented and a possible link between sorption hysteresis and kinetic properties is examined. Although the specific example of cellulose is given here, the concepts presented herein have a very wide range of applicability to many natural materials that change dimensions in response to changes in atmospheric humidity. The water sorption behaviour of cellulose is dependent upon the presence of hydroxyl groups within the structure, but this is not the only factor. A critical consideration is the mechanical response of the material. It is well known that water molecules are not able to penetrate the microfibrils and that they are only able to associate with their surface and within the matrix that exists between the microfibrils, a region often referred to the 'amorphous' or 'paracrystalline' region of cellulose. Sorbed water acts as a plasticiser for the matrix macromolecules, which consequently affects the mechanical behaviour of the material. Furthermore, the mechanical properties of the materials can affect both the sorption kinetic behaviour and the sorption isotherms.

2 THE SORPTION ISOTHERM

When cellulose is exposed to a constant relative humidity (RH) at a fixed temperature, it will gradually attain a stable moisture content (MC). This occurs when an equilibrium condition is realised and for this reason it is referred to as the equilibrium moisture content (EMC). The equilibrium state is dynamic and this point is reached when the rate at which water molecules enter the material is equal to the rate at which they exit. Determining the EMC at a constant temperature over a range of RH values produces a sorption isotherm. The sorption isotherm of cellulose is characterised by having a sigmoidal form and is classified as being International Union of Pure and Applied Chemistry (IUPAC) Type 2[1]. Apart from the sigmoidal shape, the sorption isotherm of cellulose also exhibits the property of hysteresis, where the EMC values attained during adsorption are lower than those associated with desorption at a given RH, as shown in Figure 1. The term adsorption will be used throughout this chapter since the model considered herein is one in which sorbed water molecules are associated with the internal surface (i.e. the inter-microfibrillar region) of the cellulose substrate.

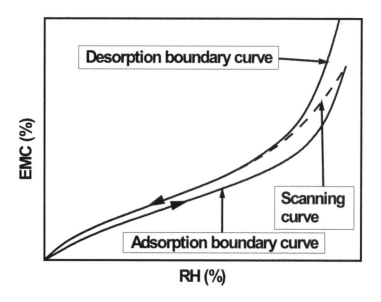

Figure 1 *Example isotherm showing the desorption boundary curve, adsorption boundary curve and the scanning curve*

The sorption isotherm consists of two boundary curves (adsorption and desorption) and a scanning curve, where desorption is initiated at a specific RH and this curve passes through the region between by the boundary curves[2]. In principle, every point within the space delineated by the boundary curves is accessible by starting a scanning curve at an appropriate point on the adsorption or desorption boundary curves. These properties have been known for over a century, yet there is still not a consensus as to why the sigmoidal isotherm is observed, nor is there a definitive explanation for hysteresis. The isotherm is generally reproducible, but this property can change if the cellulose is subjected to prolonged drying at temperatures in excess of 100 °C, which can lead to irreversible hydrogen bond formation in the matrix polymeric network[3]. It is also quite commonly observed that when never-dried cellulose is taken through the initial desorption stage that the path of this first desorption branch of the isotherm is not reproduced in any subsequent desorption experiments[4]. The reason for this is also attributed to some irreversible hydrogen bond formation (hornification) occurring in the initial drying process, irrespective of whether this takes place at elevated or ambient temperatures. The extent to which this occurs depends upon the proportion of amorphous material existing within the cellulose, some of which may be due to the presence of residual hemicellulosic polysaccharides. In practice, the determination of sorption isotherms is undertaken over a

RH range from zero to 95%, since it becomes very difficult to make accurate measurements much above this upper RH limit. There is also an increasing likelihood of a contribution from capillary condensation to the sorption process at RH values exceeding 90%[5].

Although sorption isotherms are not determined at RH values much in excess of 95%, it is quite common practice to extrapolate the adsorption isotherm line through the data points and project onto the EMC axis at 100% RH in order to determine a 'fibre saturation point' (FSP)[6]. The term FSP is used to represent the moisture content of the fibre in a theoretical condition where the cell wall moisture content is at a maximum, but there is no water present in the macrovoids of the material. In the case of wood, it is established that the FSP is a point where many of the physical properties (e.g. swelling, modulus of elasticity, modulus of rupture, impact toughness) no longer change and this is often found to be in the region of 30% MC for wood and many plant fibres[2,7]. The use of the projection of adsorption isotherms to evaluate a FSP has been criticised on a number of grounds[8]. One objection relates to the fact that the sorption isotherm is changing rapidly at this point and small errors in curve fitting lead to large errors in the projected FSP (p-FSP) value. It is also not possible to determine isotherms with any degree of accuracy much above 95% RH and it is then necessary to resort to methods such as tension plates. It is not clear how appropriate the above definition for FSP is, since many cellulosic materials are not in any case in a form that remotely resembles the cell wall from which they are derived. For example, microcrystalline cellulose (MCC) is a highly acid hydrolysed product from cellulosic fibres which is widely used as pharmaceutical excipient; its compression properties can be influenced by even very small variations in moisture content[9]. Similarly, the change in the moisture content of α-cellulose, when used in tablet formulations can affect the properties of the resulting tablets[10]. There is certainly a strong commercial imperative to understand the interaction of cellulose with moisture. This applies not only to the pharmaceutical industry, but to textiles, pulp and paper, food science and all aspects where cellulose is used industrially. Some aspects of the sorption properties of cellulose are discussed in this chapter. The sorption isotherms of native cotton linter, α-cellulose, and microcrystalline cellulose particles at 25 °C are shown in Figure 2. It is apparent that although all the samples are cellulose, that there are important differences in behaviour between them.

In this study, cotton linter exhibited a lower EMC than the α-cellulose. This is attributed to the mercerization process to obtain α-cellulose. The cotton linter contains more than 90% cellulose and a small amount of pectin and wax. The degree of crystallinity in the cellulose of cotton linter is 88%[11]. The α-cellulose is alkaline insoluble material where the pectin and wax have been removed. Compared to cotton linter, the degree of crystallinity of the α-cellulose is lower (69%) due to swelling of the crystalline region by sodium hydroxide solution[12]. Mercerization also modifies the crystallographic cell from cellulose I to cellulose II[13]. As a result, the α-cellulose exhibits an increase of the accessible area to water[14] and is able to swell to a greater extent as water is accommodated in the amorphous inter-microfibrillar matrix. The MCC exhibited a lower EMC than α-cellulose but greater than cotton linter. This can be explained by the fact that the MCC consists of aggregations of crystallites consisting of highly ordered cellulose molecules. The well-established cell wall layer network structures in the cotton linter are broken down during acid hydrolysis and the amorphous regions are dissolved, leaving mostly crystallite. The water is predominately adsorbed on the crystallite surface and in the residual amorphous regions of the MCC particles resulting in very limited swelling of the MCC[15]. The higher EMC of MCC over cotton linter may be due to a greater specific surface area of the former[16] due to the lower DP of cellulose in the crystallites.

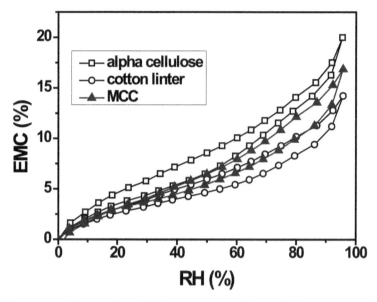

Figure 2 *Sorption isotherms of α–cellulose, microcrystalline cellulose (MCC) and cotton*

3   WATER SORPTION MODELS

A large number of models exist describing the sorption process. These can be classified as localised models, which can be broadly split into layering or cluster models[17,18] and dissolution models[19]. In localised models, the water molecules are considered to bind to specific sites in the polymeric matrix. The layering model requires the initial formation of monolayers on the cellulose internal surface (the nanopores within the inter-microfibrillar matrix), with the subsequent creation of multiple layers as the sorption process proceeds. Indeed, there is a significant component of the literature which reports on the monolayer content, often determined by the application of Brunauer Emmett Teller (BET) isotherm analysis[20]. The application of such a model can only strictly be made in the case of an inert substrate, not a swelling material, which will present an evolving surface as the sorption process takes place. This is quite apart from the consideration that the BET theory as originally formulated was applicable for gases that did not possess large dipole moments. Nonetheless, BET analyses are commonly made with water vapour sorption isotherms and BET surface areas are reported from these analyses. A modification to the BET model has led to the development of the Guggenheim-Anderson-de Boer model and its variants[21]. In dissolution models a macroscopic approach is adopted where there is mixing between macromolecules and water molecules, examples include the Hailwood and Horrobin (HH) model which was originally developed to explain the sorption isotherm of cotton[22] and the Flory-Huggins model[23]. The HH model is based upon consideration of the thermodynamics of the system where a sorbent is in equilibrium with the atmospheric moisture. The analysis considers the equilibrium existing between the solid dry polymer of the sorbent, hydrated polymer and 'dissolved' water. The hydrated polymer component is water that is strongly bound to the sorption sites of the polymeric constituents of the substrate, whereas the dissolved water is water that is not so strongly bound but is still located within the nanopores of the material. These two types of water can be loosely interpreted as monolayer and polylayer water analogously to the BET model, although the concept of a monolayer existing with the complex, dynamic geometry of the cellulose matrix internal surface is unrealistic. An HH analysis of the sorption isotherms of the three celluloses is shown in Figure 3. The sum of the two types of sorbed water produces a sigmoidal curve that very closely fits the data for the MCC and cotton linter samples, except at the upper end of the hygroscopic range where it is possible that there is an additional contribution from capillary condensation. With the α–cellulose, the fit is not so accurate since the adsorption curve does not display a sigmoidal relationship. It is important to note that the HH model cannot be used to analyse the desorption isotherm curve, since this is composed of both a boundary and a scanning curve.

4 HYSTERESIS

A number of explanations have been given in the scientific literature to explain hysteresis, but these have been unsatisfactory for a variety of reasons[5]. Early explanations relied upon the assumption that hysteresis merely represented the failure to achieve 'true' equilibrium in the experiment and that if a sufficient amount of time was allowed then the equilibrium state would be achieved. It has been shown in numerous studies that this is not the case and that hysteresis is a true phenomenon and not an experimental artefact. Other explanations have considered the nature of the adsorption process occurring onto the surfaces of the inter-microfibrillar nanopores, resulting in a step-wise filling which continues until the nanopore is filled with liquid water. Desorption then occurs from the surface meniscus of the water in the nanopore, giving rise to hysteresis. However, the concept of the existence of a meniscus in liquid water in these nanopores, which are of the order of several nm in diameter, is problematical, making such an explanation highly suspect. It has been argued that hysteresis arises due to the formation of irreversible hydrogen bonds during the first desorption cycle, which (as noted previously) can certainly explain why the first, compared to subsequent desorption boundary curves do not follow the same path, but not why the adsorption and desorption boundary curves follow different (but reproducible) paths on subsequent sorption cycles. Another model relies upon consideration of the geometry of the nanopores in the matrix, where the diameter of the throat of such a feature is smaller than the interior, akin to an ink bottle (again relying on considerations of a meniscus). Other explanations invoke differences in contact angle between the adsorption and desorption processes. During the wetting cycle the contact angle of the water with the internal surface of the cell wall is different compared with that for desorption, where the water is in contact with an already wet surface. The problem with such an explanation is that the use of concepts related to liquid water in nanopores is unlikely to have any physical meaning; furthermore, during the adsorption process, the water is entering from a vapour state and not as a liquid. It has been suggested that hysteresis is caused by the presence of permanent gases in the cell wall, leading to incomplete wetting, but it has been shown that hysteresis is still observed even if sorption isotherms are determined in a vacuum. With this model under conditions of adsorption the sorption sites are thought to be partially masked by the presence of these gas molecules. It has also been argued that the presence of permanent gases in such samples only contributes towards hysteresis if such materials are non-swelling[24].

When water vapour enters cellulose, it causes swelling of the material. This is because the water is occupying space between the microfibrils, thereby forcing them apart (water cannot penetrate the microfibrils). This process of expansion is resisted primarily by the inter-fibrillar matrix polymers and as water is removed from the substrate, the matrix collapses to its previous configuration, in other words the nanopores are not permanent but transient. The cell wall of plant fibres and many other natural materials (e.g. cellulose) can

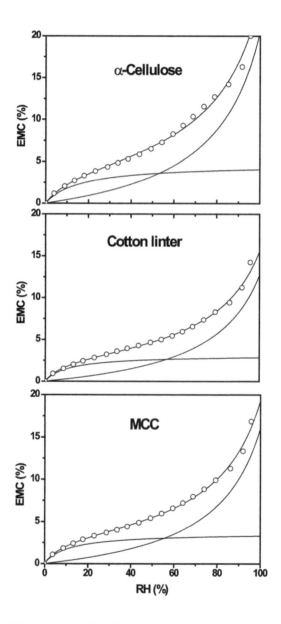

Figure 3 *Hailwood Horrobin model of the sorption isotherm showing the contributions from dissolved water and water of hydration*

be considered to be a swelling gel[25]. The sorption behaviour of natural fibres has been reported upon recently, where a model for sorption hysteresis was presented based upon the micromechanical behaviour of the cell wall matrix during the adsorption or desorption process[5]. This model was originally developed to describe sorption in glassy polymers and subsequently applied to explain sorption hysteresis in humic soils[26]. Essentially, the process of adsorption results in the creation and expansion of nanopores within the matrix, whereas desorption leads to their collapse. This process is inelastic on the time scale of molecular diffusion and as a consequence the adsorption and desorption processes take place in a material that is in different states. This phenomenon is observed in glassy polymers below the glass transition temperature,[5,27] T_g. The glassy state is defined as being one in which the molecular scale nanopores are embedded in a matrix which is unable to fully relax to a thermodynamic equilibrium state due to the stiffness of the matrix macromolecules. It is therefore logical to consider that there is a link between the matrix mechanical properties and the hysteresis effect. There are notable differences in hysteresis between the three celluloses considered in this study. Both the cotton linter and the MCC exhibit greater hysteresis at the top end of the hygroscopic range, but the α–cellulose shows a reduction in hysteresis at the RH increases. Although perhaps harder to see in Figure 2, this is better illustrated in Figure 4 which shows a comparison of the hysteresis between the adsorption and desorption isotherms (obtained by subtracting the EMC of adsorption from the EMC of desorption) for the three celluloses illustrated in this chapter.

The total hysteresis of the three celluloses was characterized by calculating the mathematical area of the isotherm loop[28]. The α-cellulose linter showed the largest isotherm loop area, which means the greatest overall sorption hysteresis. Comparatively, the MCC exhibited the lowest total hysteresis, with cotton linter in-between. In this study, as presented in an earlier section, the amorphous content of α-cellulose is the highest compared to cotton linter (which possesses a native cell wall multilayer network structure)

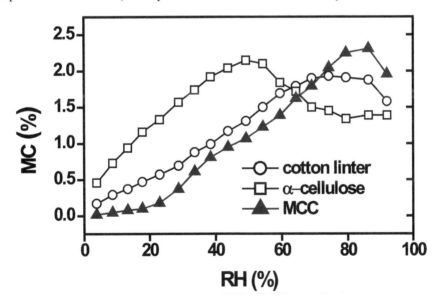

Figure 4 *Hysteresis between the adsorption and desorption isotherm loops (obtained by subtracting the EMC of adsorption from the EMC of desorption at a given RH) for microcrystalline cellulose (MCC), α-cellulose and cotton*

and MCC (essentially comprising a rigid impenetrable crystallite structure resulting from acid hydrolysis). Therefore, the extent of adsorption swelling of α-cellulose would be greater than with the cotton linter, or MCC. A further point is to be noted when comparing the hysteresis between the different cellulose samples.

The hysteresis exhibited by the α–cellulose is greater towards the lower end of the hygroscopic range, in contrast with the cotton linters and MCC. Differences in sorption hysteresis between cellulose samples of varying crystallinity have been noted before[29] where, as found herein, reduced levels of crystallinity were associated with higher levels of hysteresis in the lower half of the hygroscopic range. Finally, the T_g of the cellulose matrix will decrease as the RH of the experiment and hence moisture content of the substance increases[30,31]. At a moisture content (MC) of 15%, the T_g of ball milled eucalyptus pulp cellulose was around 65 °C, rising to 150 °C at 5% MC[30]. There would therefore be an expected decrease in hysteresis as the MC of the cellulose increases, which is what is observed with α–cellulose, but not with cotton linter or MCC. However, ball milled cellulose has a low level of crystallinity and the behaviour of the matrix in the cotton linter or MCC may not be comparable.

5 WATER ADSORPTION AND CELLULOSE SWELLING

As noted in the previous section, the adsorption of moisture into the cellulose leads to swelling of the material. The diffusion of water molecules into the matrix that envelops the microfibrils requires the breaking of hydrogen bonding networks with the creation of nanopores[28]. It is thought that the initial water molecules entering the cell wall are relatively tightly bound to the primary sorption sites (OH groups), but as the adsorption process continues, the water molecules that diffuse into the substrate nanopores are somewhat less constrained (comparatively) in their translational freedom, although this is not to say that the situation in any way resembles that of liquid water[32]. As the cellulose swells, this creates elastic strain in the matrix. The sorbed water molecules therefore perform work upon the cell wall and during the reverse process this strain energy is released as water molecules exit. However, the deformation process is not perfectly elastic since hysteresis occurs in the sorption isotherm. This is further discussed below.

6 KINETICS OF WATER SORPTION

Many models describing the kinetics of water sorption processes within natural materials have often invoked Fick's Law[19]. One of the confounding factors when attempting to model sorption kinetic behaviour has been the difficulty of obtaining sufficiently accurate data sets; an essential requirement for reliable curve-fitting. However, with the commercial development of dynamic vapour sorption (DVS) equipment over the past decade, it has now become a routine matter to obtain extremely accurate experimental kinetic data. Using a DVS apparatus, it has been clearly demonstrated that the sorption kinetic behaviour of regenerated cellulose fibres[33-35] cotton linter, α–cellulose[28] and MCC[28,36] is actually very accurately described by what is termed a parallel exponential kinetics (PEK) model, which has the mathematical form shown in Equation 1:

$$MC = MC_0 + MC_1[1 - \exp(-t/t_1)] + MC_2[1 - \exp(-t/t_2)] \tag{1}$$

where MC is the moisture content at time t of exposure of the sample to a constant RH, MC_0 is the moisture content of the sample at time zero. The sorption kinetic curve is composed of two exponential terms which represent a fast and slow process having characteristic times of t_1 and t_2 respectively. The terms MC_1 and MC_2 are the moisture contents at infinite time associated with the fast and slow processes respectively, with the sum of $MC_0+MC_1+MC_2$ being equal to the EMC. Although not yet proven, it is likely that the PEK model can be applied to all swelling gel materials, with the proviso that the sample size (or more correctly the volume to surface ratio) is small, although what the limiting ratio is has not been determined. An example sorption kinetics curve is shown in Figure 5 for cotton fibre, with the PEK fit to the data given and the fast and slow adsorption curves also reproduced.

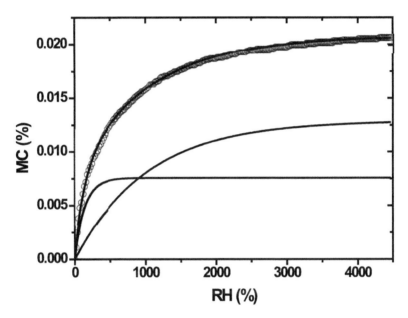

Figure 5 *Sorption kinetic curves of cotton deconvoluted into fast and slow components according to the parallel exponential kinetics (PEK) model*

It is not known what these two processes represent, but they have been previously attributed to the existence of 'fast' and 'slow' sorption sites in the material[37]. The fast sites have been attributed to the sorption of water molecules associated with the external surface and amorphous regions of the cellulose and the slow sites to indirect sorption on the inner surface and crystallites[33]. By cumulatively adding the MC_1 or MC_2 values for adsorption or desorption cycles, it is possible to construct pseudo-isotherms associated with the fast and slow processes and consequently determine what the contribution of the two processes is to the hysteresis effect. This is shown for the three forms of cellulose in Figure 6.

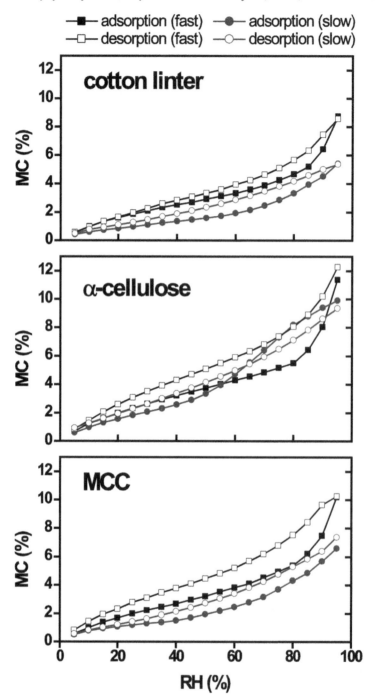

Figure 6 *Pseudo isotherms for the fast and slow processes shown for microcrystalline cellulose (MCC), α-cellulose and cotton*

With the pseudo-isotherms shown in this figure, the behaviour of the curves is not simple. With both the cotton linter and the MCC, the two sets of curves are more or less replicas of the sorption isotherm and the hysteresis has contributions from both the fast and slow processes. With the α–cellulose, the situation is much more complex and the fast process is responsible for most of the hysteresis observed in the sample; whereas the decrease in hysteresis at the upper end of the hygroscopic range is due entirely to the behaviour of the slow adsorption pseudo-isotherm line which crosses the slow desorption line at 55% RH resulting in negative hysteresis. It is not known why this occurs.

7 WATER SORPTION AND MECHANICAL BEHAVIOUR

It is well known that the modulus of elasticity and strength of wood[38] and paper[39] decrease as the moisture content is raised up to the fibre saturation point, whereas the toughness of the material increases with MC. Materials containing a high content of amorphous polysaccharide material would be expected to exhibit reductions in tensile strength and modulus from consideration of the properties of the isolated material[40]. Such systems can be considered to act as viscoelastic gels. The mechanical behaviour of such systems can be modelled as a combination of elastic and plastic deformations, as is well known when examining the mechanical behaviour of wood[41]. Since water is able to plasticise the cell wall matrix substance, it also follows that the static and dynamic mechanical properties of the material will be modified accordingly[2].

The mechanical behaviour of cellulosic materials is modified by the presence of adsorbed moisture, it is also possible to posit that the sorption behaviour is determined by the mechanical properties. It has recently been argued that the sorption kinetics of these gel-like materials is rate limited by the rate of swelling of the sorbent and that the kinetic process is in fact controlled by the matrix polymeric relaxation processes[28,42]. Fickian diffusion is an ideal case of penetrant transport, where the transport of the sorbed molecules is not affected by polymer chain rearrangement processes[19]. Whether a deviation from Fickian diffusion occurs depends upon whether the polymer relaxation process is slower than the rate of diffusion (i.e. the rate limiting step). The PEK model can therefore be interpreted as representing polymer chain relaxation as being the rate limiting step controlling the rate of diffusion of penetrant molecules. Based on this premise, the PEK model has been further interpreted in terms of two Kelvin-Voigt (K-V) viscoelastic elements acting in series (Figure 7) with E_1, E_2 being the moduli associated with the fast and slow processes respectively and η_1 and η_2 being the equivalent matrix viscosities[28].

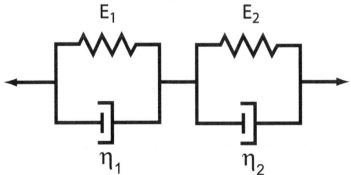

Figure 7 *Two Kelvin-Voigt (K-V) elements in series*

The two elements represent the fast and slow kinetics processes. The kinetic response of a K-V element when subjected to an instantaneous stress increase (σ_0) is as shown in Equation 2 [43]:

$$\varepsilon = (\sigma_0 / E)[1 - \exp(-t/\varphi)] \tag{2}$$

where ε is the strain at time t, E is the elastic modulus and φ is a time constant which is defined as the ratio η/E, with η the viscosity. In the case of a plant fibre subjected to a change in RH, there is a change in the swelling pressure (Π –equivalent to σ_0) exerted within the cell wall when the atmospheric water vapour pressure is raised from an initial value p_i to a final value p_f given by Equation 3 [44]:

$$\Pi = - (\rho/M)RT.\ln(p_i/p_f) \tag{3}$$

where ρ is the density and M the molecular weight of water, R is the gas constant and T is the isotherm temperature in kelvin. In the model described herein, the strain of the system is assumed to be equivalent to the volume change of the cell wall as a result of water vapour adsorption or desorption. This volume change is further assumed to be linearly related to the change in the mass fraction of the water present in the cell wall.

The adsorbed water vapour molecules exert a pressure within the cellulose leading to dimensional change, which is equivalent to the extension of the spring in the K-V model. This expansion/extension represents work and results in an increase in the free energy of the system. Expansion will continue until the free energy of the system is equal to the free energy of the water vapour molecules in the atmosphere. The spring modulus (E_1, E_2) therefore defines the water content of the system at infinite time (MC_1, MC_2) as shown in Equation 4. The fact that there is a relationship between the EMC and modulus was established over fifty years ago[45]:

$$MC_1 = \sigma_0/E_1; \quad MC_2 = \sigma_0/E_2 \tag{4}$$

The rate at which water molecules are adsorbed or desorbed by the system is a function of the viscosity of the dashpot in the model. This viscosity is in turn related to the micro-Brownian motion of the matrix macromolecular network. The more rapidly the matrix is able to deform, the faster the rate of water ingress or egress into or out of the cell wall. The rate of local deformation is related to the energy barrier associated with the local relaxation process, but also to whether there is sufficient free volume to allow these proximal relaxation processes to take place. In glassy solids below the T_g there is insufficient free volume to allow a local relaxation to take place without the cooperative motion of adjacent relaxors (a relaxor is defined as the smallest molecular segment of relaxation in each polymeric unit)[46]. This gives rise to the concept of cooperative domains within the matrix[46-48]. As the glass transition temperature is approached, the domain size decreases until T_g is reached. At this point the domain contains only one relaxor and there is sufficient free volume to allow for relaxation without the cooperation of neighbours. When this occurs, matrix relaxation becomes instantaneous and the behaviour is perfectly elastic. Hysteresis is no longer observed at this point. The presence of sorbed water molecules in the matrix can lead to a reduction of the T_g because they act as a plasticizer[31]. Consequently, hysteresis can decrease or even disappear entirely as the MC of the sample increases. The cooperative relaxation processes give rise to negative entropies of activation under conditions of adsorption and desorption[49].

Results from the K-V interpretation of the PEK data for x and y are given in Figure 8. By applying the sorption kinetic parameters to the Kelvin-Voigt model representing the viscoelastic response of the matrix, it is possible to derive values for the matrix modulus and viscosity. This has been done for the matrix modulus associated with the fast process (E_1) and slow process (E_2) and these data are presented in Figure 8. Values for the fast and slow moduli are of the order of 15 to 30 GPa at the low cellulose moisture contents associated with the bottom end of the hygroscopic range. These moduli decrease to very low values at the higher RH range, as would be predicted for a matrix that is being plasticized by the presence of sorbed water.

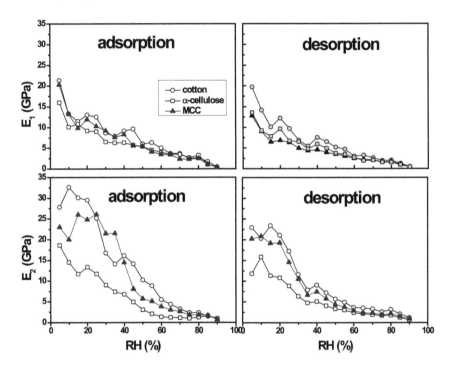

Figure 8 *Moduli associated with the fast and slow processes calculated from the PEK parameters and applying the K-V model*

There are differences in the moduli associated with the slow process between the α–cellulose and the other two celluloses in this study, especially below 40 % RH. Given that the expansion and contraction of the cellulose takes place within the inter-microfibrillar matrix, it seems reasonable to assume that the modulus values obtained are dominated by those associated with the matrix polymeric material (amorphous or paracrystalline cellulose). Unfortunately, there is little in the way of reliable direct measurement of the mechanical properties of these substances in the scientific literature.

Although Salmén[40] quotes axial microfibrillar modulus values of 134 GPa, the off-axis modulus is given as 27.2 GPa, but these values should not be affected by moisture. Thus the 'dry' modulus values calculated using the Kelvin-Voigt model in this study appear to be of the order expected given contributions both from the matrix and microfibrils, but the microfibrillar contribution is apparently absent at higher cell wall MC. The explanation for this possibly lies in consideration of the geometry of the microfibrils within the material. It may be that verification of the applicability of the Kelvin-Voigt interpretation of the sorption kinetics could be obtained by nano-indentation methods, but such experiments are difficult to perform and interpret and are not in any case directly comparable.

Differences in behaviour between the materials are also found when the matrix viscosities are examined for the fast process (η_1) and slow process (η_2) (Figure 9). For the viscosities associated with the slow process, the η_2 values are higher for desorption, but there are no major differences in the behaviour between the celluloses studied. Since the fast and slow processes have not been assigned to specific physical properties, it is not at this stage possible to comment upon the significance of such differences. What is clear is that the viscoelastic interpretation does seem to provide an insight into the sorption process and that there are important variations in behaviour that require further study. It would seem that the hysteresis effect is more associated with the slow process compared with the fast process viscosity.

It has previously been argued that an appropriate model describing sorption hysteresis in plant fibres and wood is related to the matrix response during adsorption/desorption[5]. As discussed earlier, the model describes the creation of nanopores in the matrix during the adsorption step and the collapse of these nanopores during desorption[26,27,50]. However,

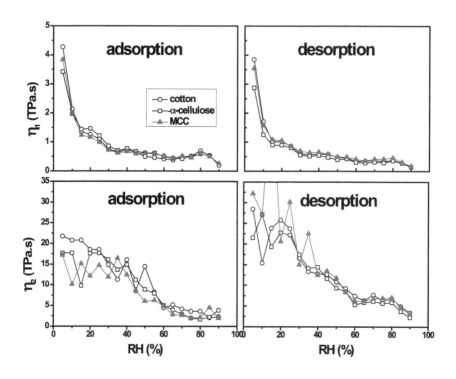

Figure 9 *Viscosities associated with the fast and slow processes calculated from the PEK parameters and applying the K-V model*

below the T_g of the matrix, the matrix is unable to respond instantaneously to the ingress or egress of water molecules because of the structural rigidity of the matrix. This results in an increased nanopore volume in the matrix during the desorption compared to the adsorption process, leading to an increased affinity for the water molecules. The time scale of the nanopore formation/annihilation processes are molecular (i.e. of the order of 10^{-10} s). By invoking the concept of cooperative relaxation in polymer systems below T_g, Matsuoka[46] is able to show that relaxation times of the order of minutes can be observed as the size of the cooperative domain increases (i.e. as the temperature decreases). This suggests that there should be a link between the hysteresis effect and the relaxation times observed in the sorption process.

10 CONCLUSIONS

The water sorption behaviour of cellulose is an important property that both influences and is controlled by the material's mechanical behaviour. It is proposed herein that there is a link between the mechanical properties of the fibre cell wall, the sorption kinetics and the property of hysteresis. The sorption kinetic behaviour can be very accurately described using the parallel exponential kinetics model, whereas Fickian behaviour is not observed. It is proposed that the sorption kinetics behaviour is described by consideration of the viscoelastic relaxation properties of the matrix. In addition, the matrix polymer relaxation behaviour can be invoked as a means of explaining hysteresis. This then suggests a link between mechanical properties and water sorption behaviour that should prove to be a promising line of research for the future.

Acknowledgement

Professor Hill acknowledges the support of the Scottish Funding Council for funding of the Joint Research Institute in Civil and Environmental Engineering, which is part of the Edinburgh Research Partnership in Engineering and Mathematics.

References

1 S.J. Gregg and K.S.W. Sing, *Adsorption, Surface Area and Porosity*, 2nd Edition, Academic Press, London, 1982.
2 W.E. Morton and J.W.S. Hearle, *Physical Properties of Textile Fibres*, 4th Edition, 2008, CRC Press, Boca Raton.
3 S. Zervos, *Cellulose*, 2007, **14**, 375.
4 P. Ingram, D.K. Woods, A. Peterlin and J.L. Williams, *Text. Res. J.*, 1974, **44**, 96.
5 C.A.S. Hill, A. Norton and G. Newman. *J. Appl. Polym. Sci.*, 2009, **112**, 1524.
6 C. Skaar, *Water in Wood*, Syracruse University Press, New York, 1972.
7 J.F. Siau, *Transport Properties in Wood*, Springer-Verlag, Berlin, 1984.
8 A.J. Stamm, *Wood Sci.*, 1971, **4**, 114.
9 F. Doelker, *Drug. Dev. Ind. Pharm.*, 1993, **19**, 2399.
10 A.O. Okhamafe, A.C. Igboechi, C.E. Ubrufih, B.O. Akinyemi and M.O. Ighalo, *Pharm. World J.*, 1992, **1**, 11.
11 C. Baillie. *Green Composites: Polymer Composites and the Environment*. Woodhead Publishing Limited, New York, 2004.
12 H.A. Krässig, *Cellulose: Structure, Accessibility and Reactivity*. Gordon and Breach Science Publishers, Switzerland, 1993.

13 K. Stana-Kleinschek, S. Strnad and V. Ribitsch, *Polym. Eng. Sci.* 1993, **39**, 1412.
14 J.H. Wakelin, H.S. Virgin and E. Crystal, *J. Appl. Phys.*, 1959, **30**, 1654.
15 M.A. Strømme, A. Mihranyan, R. Ek and G.A. Niklasson,. *J. Phys Chem. B*, 2003, **107**, 14378.
16 J.F. Kennedy, G.O. Phillips, D.J. Wedlock, P.A. Williams, *Cellulose and its Derivatives: Chemistry, Biochemistry and Applications*. Ellis Horwood Limited, Chichester, 1985.
17 A. Venkateswaram, *Chem. Revs.* 1970, **70**, 619.
18 W. Simpson, *Wood Fiber Sci.*, 1980, **12**, 183.
19 G.K. van der Wel and O.C.G. Adan, *Progr. Org. Coat.*, 1999, **37**, 1.
20 S. Brunauer, P.H. Emmett and E. Teller, *J. Am. Chem. Soc.*, 1938, **60**, 309.
21 J. Chirife, O. Timmermann, H.A. Inglesias and R. Boquet, *J. Food Eng.*, 1992, **15**, 75.
22 A.J. Hailwood and S. Horrobin, *J. Chem. Soc. Faraday Trans.*, 1943, **42B**, 84.
23 P.J. Flory, *Principles of Polymer Chemistry*, Cornell University Press, New York, 1953.
24 A.J. Stamm, *Wood and Cellulosic Chemistry*. Ronald Press, New York, 1964.
25 W.W. Barkas, *Trans. Faraday Soc.* 1942, **38**, 194.
26 Y. Lu, and J.J. Pignatello, *J. Env. Qual.* 2005, **34**, 1063-1072.
27 Y. Lu and J.J. Pignatello, *Env. Sci. Technol.* 2002, **36**, 4553.
28 Y. Xie, C.A.S. Hill, Z. Jalaludin and D. Sun, *Cellulose*, 2011, **18**, 517.
29 A. Mihranyan, A.P. Llagostera, R. Karmhag, M. Strømme and R. Ek, *Int. J. Pharm.*, 2004, **269**, 433.
30 S.S. Paes, S. Sun, W. MacNaughtan, R. Ibbett, J. Ganster, T.J. Foster and J.R. Mitchell. *Cellulose*, 2010, **17**, 693.
31 D. Kilburn, J. Claude, T. Schweizer, A. Alam and J. Ubbink, *Biomacromol.*, 2005, **6**, 864.
32 J.C.F. Walker, *Primary Wood Processing: Principles and Practice*, 2nd edition, Springer, Dordrecht, 2006.
33 S. Okubayashi, U.J. Griesser and T. Bechtold, *Carb. Polym.*, 2004, **58**, 293.
34 S. Okubayashi, U.J. Griesser and T. Bechtold, *J. Appl. Polym. Sci.*, 2005, **97**, 1621.
35 S. Okubayashi, U.J. Griesser and T. Bechtold, *Cellulose*, 2005, **12**, 403.
36 K. Kachrimanis, M.F. Noisternig, U.J. Griesser and S. Malamataris, *Eur. J. Pharm. Biopharm.*, 2006, **64**, 307.
37 R. Kohler, R. Dück, B. Ausperger and R. Alex, *Compos. Int.* 2003, **10**, 255.
38 J.M. Dinwoodie, *Timber: Its Nature and Behaviour*, 2nd Edition, E. and F.N. Spon, London, 2000.
39 H.W. Haslach, *Mech. Time Dependent Mat.*, 1995, **4**, 169.
40 L. Salmén, *C. R. Biol.*, 2004, **327**, 873.
41 J. Passard, and P. Perré, *Ann. For. Sci.*, 2005, **62**, 823.
42 C.A.S. Hill and Y. Xie. *J. Mater. Sci.* 2011, **46**, 3738.
43 H.A. Barnes, J.F. Hutton and K. Walters, *An Introduction to Rheology*, Elsevier, Amsterdam, 1989.
44 K. Krabbenhoft and L. Damkilde, *Matérieux et Constr.*, 2004, **37**, 615.
45 R. Meredith, *J. Text. Inst. Trans.*, 1957 **48**, T163-T174.
46 S. Matsuoka, *Relaxation Phenomena in Polymers*. Hanser, New York, 1992.
47 S. Matsuoka and A. Hale, *J. Appl. Polym. Sci.*, 1997, **64**, 77.
48 A. Bartolotta, G. Carini, G. Carini, G. Di Marco, G. Tripodo, *Macromolecules*, 2010, **43**, 4798.
49 C.A.S. Hill, B.A. Keating, Z. Jalaludin, E. Mahrdt, *Holzforschung*, 2012, **66**, 35.
50 J.S. Vrentas and C.M. Vrentas, *Macromolecules* 1996, **29**, 4391.

12

LIGNIN BIOSYNTHESIS AND LIGNIN MANIPULATION

P. Daly, M. Maluk, M. Zwirek and *C. Halpin

Division of Plant Sciences, College of Life Sciences, University of Dundee at the James Hutton Institute, Invergowrie, Dundee DD2 5DA, Scotland, UK
*c.halpin@dundee.ac.uk

1 INTRODUCTION

Lignin is critical to normal plant health and development but is sometimes less useful from an industrial perspective where it is often seen as a troublesome component of plant biomass. For the past 20 years, research has explored the opportunities for manipulating lignin by genetic engineering to improve biomass for industrial or agricultural processes. Initial work focussed on improving wood for papermaking by reducing lignin content, or making it easier to extract during pulping, so that the costs and environmental impact of isolating relatively clean cellulose fibres were reduced. The prospect of making forages more digestible by manipulating lignin to enable greater release of the energy stored in wall carbohydrates, has also been explored. Today, there is a major push towards manipulating lignin to allow more efficient biofuel production from plant biomass by allowing enzymes easier access to the polysaccharide components of cell walls that can be used as substrates for fermentation into fuels.

In all of these processes, the reasons that lignin is so problematic are directly related to the reasons that it evolved in the first place and its function in plants. Lignin is a hydrophobic polymer that encrusts the other components of secondary cell walls particularly in the vascular system, in order to waterproof them and confer structural rigidity. Over evolutionary timescales, the development of lignin was critical in enabling plants to move onto land and to allow the generation of tall structures like trees that could still conduct water efficiently from root to canopy top. Lignin also plays a role, not always well-defined, in plant defence and is often deposited at infection or wound sites to prevent or limit pathogen ingress and spread, as few organisms have evolved the capability of degrading lignin. Lignin is therefore one of the most slowly decomposing components of dead vegetation and plays a significant role in the biogeochemical carbon cycle.

2 LIGNIN STRUCTURE

The difficulties in degrading lignin arise from the polymer's complicated structure of phenylpropanoid units connected by a variety of chemical bonds produced during random free radical polymerisation of the lignin monomers. Although many bonds are labile, principally the β-O-4 ether bonds, many resistant carbon-carbon bonds (β-5), biphenyl bonds (5-5) and condensed biphenyl ether linkages (4-O-5) are also present (Figure 1). The three major lignin monomers (monolignols), i.e. p-coumaryl alcohol,   coniferyl alcohol   and   sinapyl alcohol, are methoxylated to various degrees and give rise to what are know as *p*-hydroxyphenyl (H), guaiacyl (G) and syringyl (S) lignin units respectively. Guaiacyl units have two sites for coupling into the growing polymer while syringyl units have just one. Lignin rich in G units will therefore have more resistant carbon-carbon (β-5) bonds than lignins having more S units where the C_5 is not available for coupling[1]. The relatively less prevalent H units also impact significantly on lignin structure and chemical properties by introducing resistant carbon-carbon linkages[2]. Covalent links to hemicellulose are also present, cross-linking lignin within the entire cell wall network. The lack of a repeating structure in lignin has hampered microbes from developing a simple enzyme system for degrading it although a small number of fungi and bacteria secrete lignin-degrading enzymes (ligninases).

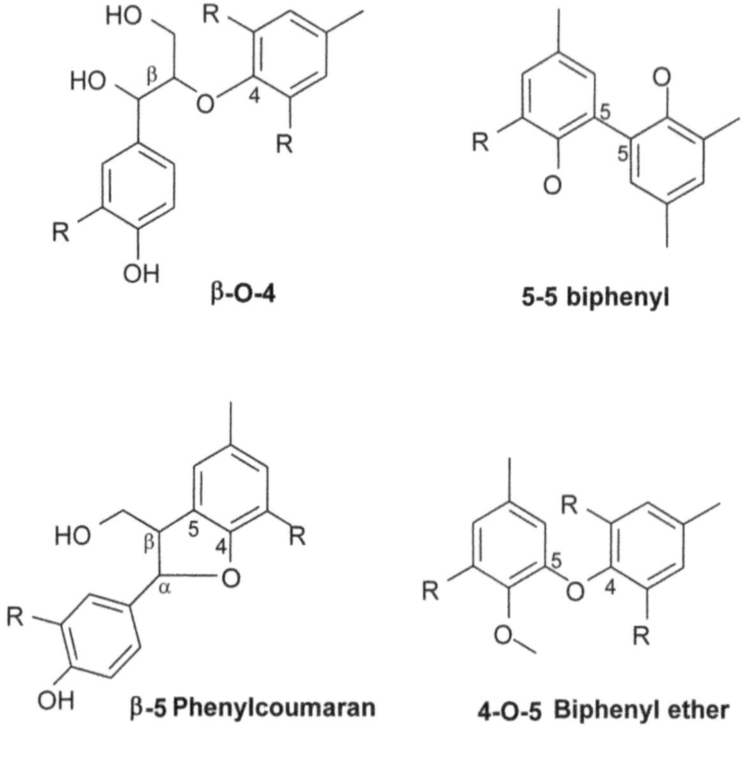

Figure 1 *Major bonds in lignin*

Efforts to manipulate lignin must achieve the required improved properties for industrial processing (i.e. increasing access to the complex carbohydrate structures in the cell wall) without interfering with the polymer's important functions in the plant. This may seem like a very difficult prospect but in fact huge variability exists in nature between plant species for both the amount and composition of lignin, supporting the idea that some manipulation should be possible. Indeed the monomer composition of lignin can differ with plant variety, tissue, cell type and even developmental stage[3], so it does not have a fixed composition and structure.

3 PLANT BIOMASS AND SECOND GENERATION BIOFUELS

The value of being able to produce plant biomass with compositions tailored to different industrial applications could be enormous, as illustrated by the research focussing on facilitating the production of second generation biofuels. Such fuels have the capacity to replace finite fossil fuels with renewable fuel sources and reduce dependency on imports while lowering CO_2 emissions and thereby mitigating climate change. While we have many renewable options for producing electrical energy (water, wind, etc) we have fewer short- to mid-term options for replacing current liquid transportation fuels. Biofuels are one of the most promising alternatives and are currently produced at large scale from materials such as sugar cane or corn grain where sugars or starch are readily available. However, increasing pressure on such potential food supplies as the world population increases suggest that a more rational approach is to use the sugars from cellulose in plant biomass for biofuel production. This fuel could be produced from agricultural or forestry wastes such as straw, bagasse and wood clippings, or from dedicated biomass crops (*Miscanthus*, switchgrass, willow, poplar) grown on poorer soil. However, such cellulosic biofuels are not currently very economic to produce due to the increased complexity of the feedstock which requires extensive and costly processing and pre-treatment with heat or chemicals in order to break open the structure of cell walls and remove lignin so that enzymes can gain access to the cellulose and hydrolyse it to the sugar substrates for biofuel production. Lignin appears to inhibit several aspects of the process by acting as a physical barrier, impeding swelling of cellulose fibers, and non-specifically binding cellulase enzymes to inhibit activity[4]. Scientific research on the biosynthesis, structure and manipulation of plant cell walls and their component polymers could help to make cellulosic ethanol production sufficiently cost efficient to operate at a full commercial scale. For example, if we can reduce lignin content or alter its structure to enable less stringent pretreatments or to facilitate greater enzymatic release of sugars from cellulose, we may decrease the cost and increase the efficiency of biofuel production.

4 MANIPULATING LIGNIN BIOSYNTHESIS – TRANSGENIC APPROACHES

We can manipulate the content and composition of lignin in plants by suppressing the expression of genes involved in lignin biosynthesis using molecular silencing techniques such as RNA interference (RNAi). This is a widely used genetic transformation technique which serves as a tool to study gene function in plants through genetic engineering. RNAi involves stable transformation with a gene construct that, when expressed, produces a small double stranded RNA homologous to the target gene that subsequently marks that gene's transcripts for degradation[5]. At least 10 potential target genes are suggested by the known enzymes of the phenylpropanoid pathway that are directly involved in lignin monomer production. All of these genes have been manipulated in transgenic or mutant plants and the consequent modifications to lignin have been described. A summary compiled from numerous studies of the consequences of changing the expression of monolignol biosynthesis genes on lignin amount and composition is given in Vanholme et al[6]. The exact changes to lignin content or lignin structure that is achieved by manipulating the expression of each individual gene depends on the specific role of that gene in the pathway. Thus, caffeic acid O-methyltransferase (COMT) is involved in the conversion of G unit precursors into S unit precursors, adding a second methyl group to 5-hydroxyconiferaldehyde to make sinapaldehyde. When COMT is suppressed it inhibits flux into S lignin biosynthesis; the S:G ratio of the polymer is reduced and unusual 5-OH G units are incorporated into lignin[7-11]. By contrast when cinnamyl alcohol dehydrogenase (CAD) is suppressed, the conversion of cinnamaldehydes into cinnamyl alcohol is reduced and the aldehydes get incorporated into lignin in place of the normal alcohols[12-16]. This results in a lignin that is more labile to alkali treatment with more free phenolic end groups. Cinnamoyl Co-A reductase (CCR) acts before the branchpoint to G and S lignin units and suppression of this gene results in reduced lignin content[17-18]. While some of the plants produced by such lignin manipulation do not grow to normal heights, others appear perfectly normal. For example, CCR-suppressed plants can be stunted and have dark green leaves with curling edges but CAD- or COMT-suppressed plants grow to normal heights and are indistinguishable from non-modified plants. This is true even with modified trees growing in the field. Poplars suppressed in these genes were grown for 4 years in a controlled field-trial and, despite the types of modification to lignin described above, the transgenic trees reached the same height and girth as non-modified control trees and had no detectable change in appearance[19].

Plants modified in this way can have very useful properties for specific applications, for example, for the CAD-suppressed poplars, it was easier to delignify the cellulose fibres during pulping to make paper, while both the CAD- and COMT-suppressed tobacco allowed greater enzymatic release of simple sugars from cell wall polysaccharides after an acid pre-treatment[20]. Decomposition of the roots of the poplar trees was also altered with the transgenic roots decomposing slightly but significantly faster than wild type roots[19].

Although much of the work on transgenic lignin manipulation has been performed in dicots (e.g. tobacco, Arabidopsis, poplar), similar methods are beginning to be applied to monocots useful for sustainable energy production such as dedicated bioenergy/biofuel grasses (switchgrass, sorghum, *Miscanthus*) and agricultural crops with straw or stover wastes (e.g. barley, maize). For example, switchgrass is a USA native perennial warm-season C_4 grass, which has been considered as a good candidate for lignocellulose feedstock production[21]. Successful transgenic modification of switchgrass by suppression of COMT showed increased sugar yield and enabled the production of 14-28% more ethanol per gram of cellulose depending on pre-treatment conditions[22]. Although these plants showed normal growth and development, when COMT was similarly down-regulated in perennial ryegrass, reductions in plant vigour and increased susceptibility to a pathogen were observed[23]. The biggest challenge to this kind of genetic modification therefore continues to be the problem of achieving effective manipulation without causing adverse effects on development or biomass yield.

5 MANIPULATING LIGNIN BIOSYNTHESIS – NATURAL VARIATION

Although the changes to lignin described here were introduced into plants by genetic manipulation, the same kinds of modification exist in nature but are generally hard to detect unless an obvious phenotype change occurs. For a couple of lignin biosynthesis genes this is the case. When either COMT or CAD are knocked out in maize or sorghum, the plants have a visible brown midrib in the leaves and are known as *brown-midrib* mutants. These mutations can be useful - *brown-midrib* plants have lower lignin content and increased forage quality due to improved digestibility. Two separate studies also demonstrate an increase in saccharification of approximately 20% for the sorghum *bmr12* *brown-midrib* mutant (a COMT mutant) using acid pretreatments[24-25]. It would be useful to be able to identify similar mutants in other species. Using the red-brown lignin phenotype as an identifier, we have been able to identify CAD mutants in barley[26] and others have found similar mutants in rice[27] and loblolly pine[28]. These barley mutants have the same useful changes to lignin and brown-midrib (or orange lemma) barley has improved saccharification[26] while the loblolly pine CAD mutant is easier to pulp[29-30]. Thus, using the transgenic route we are merely reproducing modifications to lignin that are within the normal range found in nature but that are difficult to identify in conventionally bred crop gene pools.

A second approach to discovering such useful natural variations in lignin and to exploiting them to discover the underlying genes that are influencing these cell wall traits, is to survey large populations of different varieties of a crop for the desired properties. In one study of this type, natural variation in lignin structure was evaluated in an undomesticated population of 1,100 *Populous trichocarpa* trees[31]. Significant variation for both lignin content and S/G ratios were identified which had impacts on saccharification yields. We are performing a similar type of analysis in barley using the amount of simple sugars that can be released from cell wall polymers (saccharification) in straw from hundreds of different varieties as a primary screen[26]. This is done using a semi-automated high-throughput robotic platform[20]. Although we expect lignin to be one of the cell wall properties that influence this trait, it may not be the only one and, using this system, we may also learn something about the natural variation to hydrolysis susceptibility of the carbohydrate polymers themselves. Our ultimate aims are two-fold: to identify the barley varieties that might produce straw most suitable for biofuel production, and, to identify the underlying genetic bases determining the ease or difficulty of saccharification. By genotyping the same barley populations and then correlating the saccharification data with the presence or absence of different genetic markers in each variety, we can identify regions in the barley genome that are controlling saccharification and we can subsequently work towards identifying the genes in those regions. This approach is known as association genetics or GWAS (Genome-Wide Association Study). Identified genes and specific gene alleles can then become targets for marker-assisted approaches to crop improvement through conventional breeding or could be targets for genetic manipulation.

6 PERSPECTIVES

Research to date has illustrated the possibilities for manipulating lignin *in planta* to maximise the efficiency of plant biomass processing during industrial operations like paper-making and biofuel production. Continuing research focus is this area will no doubt eventually result in the production of biomass crops or wastes that are better suited to processing than current varieties. For current agricultural crops, similar research efforts to improve yields will enable more food (from grain) and fuel (from straw) to be produced per land area from dual purpose crops. For dedicated biomass grasses, efforts towards improving nutrient and water use efficiency may result in bioenergy crops that will tolerate less fertile soils leaving the best agricultural land for food production. Maximising the exploitation of complex cell wall polysaccharides for food and biofuel products must be balanced by more inventive ways of using lignin itself for energy, fuel or chemical production or, indeed, for C sequestration in soils. By manipulating lignin differently in stems and roots, for example, it may be possible to produce plants where low lignin in aerial parts makes cellulose more accessible for biofuel production while higher lignin in roots retards decomposition, locking C in the soil for longer.

References

1   W. Boerjan, J. Ralph, and M. Baucher, *Ann. Rev. Plant Biol.,* 2003, **54,** 519.

2   Y. Barriere, C. Riboulet, V. Mechin, S. Maltese, M. Pichon, A. Cardinal, C. Lapierre, T. Lubberstedt, and J.-P. Martinant, *Genes, Genomes and Genomics,* 2007, **1,** 133.

3   M. Campbell and R. Sederoff, *Plant Physiol.* 1996, **110,** 3.

4   W. Vermerris, A. Saballos, G. Ejeta, N.S. Mosier, M.R. Ladisch, and N.C. Carpita *Crop Sci.* 2007, **47,** S142.

5   S.V. Wesley, C.A. Helliwell, N.A., Smith, M. Wang, D.T. Rouse, Q. Liu, P.S. Gooding, S.P. Singh, D. Abbott, P.A. Stoutjesdijk, S.P. Robinson, A.P. Gleave, A. G. Green, and P.M. Waterhouse, *Plant J.,* 2001, **27,** 581.

6   R. Vanholme, K. Morreel, J. Ralph and W. Boerjan, *Curr Opin Plant Biol.,* 2008, **11,** 278.

7   R. Atanassova, N. Favet, F. Martz, B. Chabbert, M.T. Tollier, B. Monties, B. Fritig, and M. Legrand M. *Plant J.,* 1995, **8,** 465.

8   J. Van Doorsselaere, M. Baucher, E. Chognot, B. Chabbert, M.T. Tollier, M. PetitConil, J.C. Leple, G. Pilate, D. Cornu, B. Monties, M. VanMontagu, D. Inze, W. Boerjan, and L. Jouanin, *Plant J.,* 1995, **8,** 855.

9   C .Lapierre, B. Pollet, B.M. Petit-Conil, G. Toval, J. Romero, G. Pilate, J.C. Leple, W. Boerjan, V. Ferret, V. de Nadai, and L. Jouanin, (1999). *Plant Physiol.* 1999, **119,** 153.

10 L. Jouanin, L, T. Goujon, V. de Nadai, M.T. Martin, I. Mila, C. Vallet, B. Pollet, A. Yoshinaga, B. Chabbert, M. Petit-Conil, and C. Lapierre, *Plant Physiol.,* 2000, **123,** 1363.

11 J.M. Marita, J. Ralph, R.D. Hatfield, D.J. Guo, F. Chen, and R.A. Dixon, *Phytochem.* 2003, **62,** 53.

12 C. Halpin, M.E. Knight, G.A. Foxon, M.M. Campbell, A.M. Boudet, J.J. Boon, B. Chabbert, M.-T. and W. Schuch. *Plant J.,* 1994, **6,** 339.

13 H. Kim, J. Ralph, F.C. Lu, G., Pilate, J.C. Leple, B. Pollet, B, and C. Lapierre, *J. Biol. Chem.* 2001, **277,** 47412.

14 J. Ralph, J.J. MacKay, R.D. Hatfield, D.M. O'Malley, R.W. Whetten, and R.R. Sederoff, *Science,* 1997, **277,** 235-239.

15 J. Ralph, R.D. Hatfield, J. Piquemal, N. Yahiaoui, M. Pean, C. Lapierre, and A.M. Boudet, *Proc. Natl. Acad. Sci. U.S.A.,* 1998, **95,** 12803.

16 J. Ralph, C. Lapierre, J.M. Marita, H. Kim, F.C. Lu, R.D. Hatfield, S. Ralph, C. Chapple, R. Franke, M.R. Hemm, J. Van Doorsselaere, R.R. Sederoff, D.M. O'Malley, J.T. Scott, J. J. Mackay, N. Yahiaoui, A.M. Boudet, M. Pean, G. Pilate, L. Jouanin, and W. Boerjan, *Phytochem.* 2001, **57,** 993-1003.

17 J. Piquemal, C. Lapierre, K. Myton, A. O'Connell, W. Schuch, J. Grima-Pettenati, and A.M. Boudet, *Plant J.* 1998, **13,** 71.

18 A. O'Connell, K. Holt, J. Piquemal, J. Grima-Pettenati, A.M. Boudet, B. Pollet, C. Lapierre, M. Petit-Conil, W. Schuch, and C. Halpin, *Transgenic Res.* 2002, **11**, 495.

19 G. Pilate, E. Guiney, M. Petit-Conil C, Lapierre J.-C., Leple B. Pollet I. Mila, E.A. Webster, H.G. Marstrop, D.W. Hopkins, L. Jouanin, W. Boerjan, W. Schuch, D. Cornu and C. Halpin, *Nat. Biotechnol.* 2002, **20**, 607.

20 L. Gomez, C. Whitehead, A. Barakate, C. Halpin, and S. McQueen-Mason, *Biotechnol. Biofuels,* 2010, **3,** 23

21 S.B. McLaughlin and L. Adams Kszos, *Biomass and Bioenergy,* 2005 **28,** 515.

22 C.X. Fu, J.R. Mielenz, X. R. Xiao, Y.X. Ge, C.Y. Hamilton, M. Rodriguez, F. Chen, M. Foston, A. Ragauskas, J. Bouton, R.A. Dixon, and Z.Y. Wang, *Proc. Natl. Acad. Sci. U.S.A.,* 2011, **108,** 3803.

23 Y. Tu, S. Rochfort, Z. Liu, Y. Ran, M. Griffith, P. Badenhorst, G.V. Louie, M.E. Bowman, K.F. Smith, J.P. Noel, A. Mouradov, and G. Spangenberg, G. *Plant Cell,* 2010, **22,** 3357-3373.

24 A. Saballos, W. Vermerris, L. Rivera, and G. Ejeta, *Bioenergy Res.,* 2008, 1:193.

25 B.S. Dien, G. Sarath, J.F. Pedersen, S.E. Sattler, H. Chen, D.L. Funnell-Harris, N.N. Nichols, and M.A. Cotta, *Bioenergy Res.,* 2009, **2**, 153.

26 C. Halpin, R. Shafiei, J. Kam, Y. Wilson, P. Daly, M. Maluk, M. Zwirek, S. McQueen-Mason, L. Gomez, C. Whitehead, J. Comadran, N. Uzrek and R. Waugh, in *Proceedings of the Bioten Conference on Biomass Bioenergy and Biofuels 2010,* ed. A.V. Bridgwater, CPL Press, 2011.

27 K. Zhang, Q. Qian, Z. Huang, Y. Wang, M. Li, L. Hong, D. Zeng, M. Gu, C. Chu, and Z. Cheng. *Plant Physiol.* 2006, **140**, 972.

28 J.J. MacKay, D.M. O'Malley, T. Presnell F.L. Booker, M.M. Campbell, R.W. Whetten and R.R. Sederoff, *Proc. Natl. Acad. Sci. U.S.A.* 1997, **94**, 8255.

29 D.R. Dimmel, J.J. MacKay, E.M. Althen, C.J. Parks, and R.R. Sederoff, *J. Wood Chem. Technol.* 2001, **21**, 1.

30 D.R. Dimmel, J.J. MacKay, C. Courchene, J. Kadla, J.T. Scott, D.M., O'Malley and S.E. McKeand, *J. Wood Chem. Technol.* 2002, **22,**.

31 M.H. Studer, J.D. DeMartini, M.F. Davis, R.W. Sykes, B. Davison, M. Keller, G.A. Tuskan, and C.E. Wyman, *Proc. Natl. Acad. Sci. U.S.A.,* 2011, **108**, 6300.

13

BACTERIAL DEGRADATION OF ARCHAEOLOGICAL WOOD IN ANOXIC
WATERLOGGED ENVIRONMENTS

N.B. Pedersen[1*], C.G. Björdal[2], P. Jensen[3] and C. Felby[1]

[1]University of Copenhagen, Faculty of Science, Rolighedsvej 23, 1958 Frederiksberg C, Denmark
[2]University of Gothenburg, Department of Conservation, Guldhedsgatan 5A, 413 20 Göteborg, Sweden
[3]National Museum of Denmark, Department of Conservation, I.C. Modewegsvej, 2800 Kgs. Lyngby, Denmark
*nape@life.ku.dk

1 INTRODUCTION

Wood contains the most dominant and abundant biopolymers in the global ecosystem. The main chemical components of wood are carbohydrates such as cellulose and hemicellulose and the polyphenol lignin which are all assembled into a complex cell wall matrix. On a yearly basis 80-100 billion tons of wood are produced by photosynthesis and it represents the largest input of biogenic carbon to the global carbon cycle. The biogenic carbon is returned as inorganic carbon after a complex interaction of decay mechanisms of biotic and abiotic origin. Wood decays within years to decades under aerobic conditions, mainly by the action of wood degrading fungi. In anoxic ecosystems such as bogs, marine and freshwater sediments, and waterlogged soils the decay takes place by bacteria and abiotic factors during centuries, millennia and in some extreme cases even millions of years[1,2]. The resistance of wood to biotic and abiotic decay is caused not only by the anoxic environmental conditions but also by the structure and organisation of the cell wall components; a generic property that is often termed recalcitrance[3]. Anoxic (anaerobic) waterlogged environments are large sinks in the global carbon cycle, but the decay mechanisms acting in these environments are not fully understood.

Wood has served as an important material in human culture throughout history. It has been used both for large constructions such as ships, bridges, and houses, more delicate objects such as furniture, sculptures, tools, weapons, and last but not least for energy (Figure 1). Wood from past cultures and forests have in some instances been submerged in anoxic or near anoxic environments and have therefore been preserved until the present. Compared to the huge amount of wooden objects that have been in use throughout human history only relatively few wooden archaeological objects have however been preserved. The preserved objects are well suited for studying the degradation of biogenic carbon in anoxic environments thereby providing not only archaeological and cultural knowledge but also enabling us to study the chemistry and biology of the decay mechanisms and their interactions with the biopolymers in wood.

The structure of archaeological wood at the macroscopic, microscopic, and ultrastructural level is crucial in relation to the properties of the material. The residual ultrastructure of the wood and the chemical composition of the ultrastructural components left after decay controls the behaviour of the wood. To preserve the original valuable material for future scientific work and to save important information the wood has to be dried to obtain a stabile form suitable for display and storage. However the decayed wood structure will collapse, shrink, and warp if the structure is simply air dried under un-controlled conditions. Conservation treatments used in the past and present have mainly been based on trial and error; and to a lesser extent on macroscopic and microscopic material properties. This has led to a great deal of successful preservation on the one hand[4,5] but also to serious damage of important cultural heritage objects on the other[6,7]. A deeper understanding of the residual ultrastructure and the chemical composition in waterlogged archaeological wood will be invaluable when new conservation treatments are developed.

Figure 1 *Small trading and transport vessel (Skuldelev 3) from the Viking Age, built of Danish oak in approximately 1040. The wooden ship has been preserved for more than 900 years in anoxic sediments in Roskilde Fjord, Denmark before it was excavated in 1962. The wood was soft and spongy upon excavation due to decay, but surface details such as tool marks were remarkably well preserved. Picture ©: the Viking Ship Museum in Roskilde, Denmark. Photo: Werner Karrasch*

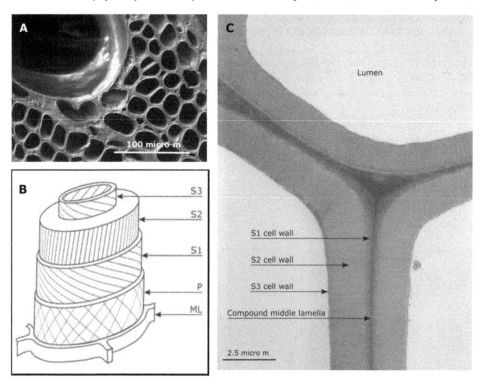

Figure 2 *Wood tissue and ultrastructure of the wood cell wall. A: Scanning electron microscopy cross section of ash (*Fraxinus excelsior *L.) wood tissue. Picture: National Museum of Denmark. B: Schematic representation of a wood cell with location of middle lamella (ML), primary wall (P), and the three secondary cell wall layers (S$_1$, S$_2$, and S$_3$). The microfibril distribution and angles in the different cell wall layers are shown. Redrawn from Fujita and Harada 2001[10]. C: Transmission electron microscopy cross section of cell wall in early wood tracheids of Norway spruce (*Picea abies *(L.) Karst) with location of compound middle lamella and the three secondary cell wall layers (S$_1$, S$_2$, and S$_3$). Picture ©: Nanna Bjerregaard Pedersen*

The decay organisms and the ultrastructural and chemical composition of archaeological waterlogged wood have been the focus of much research - but a large body of knowledge is still lacking. The chemical composition of the polymers in the residual ultrastructure as well as the reaction mechanisms of the degradation processes are only fundamentally understood. In addition the actual species of bacteria responsible for the main decay have yet to be revealed as well as the understanding of the interactions between the environmental and material factors controlling decay rates. The scope of this contribution is to present an overview of the numerous morphological and chemical composition studies performed on waterlogged archaeological wood and of the known environmental parameters that control the microbial decay of waterlogged wood in the submerging environment. This contribution has special focus on erosion bacteria as this is the most dominant decay type in waterlogged archaeological wood found in anoxic environments.

2   ORGANISATION AND STRUCTURE OF WOOD

A number of different cell types can be found in wood tissue. Softwood has a relatively simple structure of tracheid and parenchyma cells whereas hardwood has a wide range of fiores, parenchyma and vessel elements (Figure 2A). Furthermore each cell type may vary at the anatomical level where latewood, earlywood, heartwood, sapwood, and reaction wood are examples of how the cells adapt to the physiological conditions within the tree. The wood tissue consists of an axial and radial system of cells connected through a variety of pits. The pits control the transport of sap and water in the living wood; and they are the most frequent points of entry for the wood degrading organisms in dead wood[8-12]. There is a close relationship between the structure of the individual cell types and their recalcitrant properties[13]. However, all wood cell walls consist of between 65-75% carbohydrate and 25-35% lignin. The cellulose is arranged in microfibrils which are highly ordered bundles of parallel cellulose chains held together by hydrogen bonding. The microfibrils provide the structural reinforcing network of the cell wall. The cellulose is embedded in a matrix of hemicellulose and lamellar lignin. There are no covalent bonds between the cellulose polymers and the other polymers, whereas lignin and hemicellulose are linked through ester bonds known as lignin carbohydrate complexes[8-12].

The wood cell wall is organised into primary and secondary wall layers surrounded by a middle lamella holding the individual wood cells together. The primary wall is a thin cell wall located next to the middle lamella (Figure 2B and Figure 2C). It consists of cellulose fibrils arranged in thin crossing layers with high amounts of embedded pectin. The transition from the middle lamella to the primary wall is diffuse and the two layers are often termed compound middle lamella. The composition of the compound middle lamella is approximately 60% lignin, 20-30% hemicellulose and 10-15% cellulose. The secondary wall is a thick cell wall subdivided into three layers S1, S2, and S3 from the primary wall and inwards (Figure 2B and Figure 2C). Each layer has a different angle of the microfibrils (Figure 2B). The S1 layer has a microfibril orientation shifting gradually and in a clockwise direction from an S-helix (> 90°) to a Z-helix (< 90°) from the outer to the inner part of the layer. The S2 cell wall layer is the thickest layer in the cell wall accounting for 50% to 87% of the thickness of the entire cell. Approximately 60% of the secondary wall is cellulose, 15% is hemicellulose and 25% is lignin. In the S2 cell wall layer the galacto-glucomannans are physically associated more with the cellulose fibrils whereas the heteroxylans are more associated with the lignin. The microfibril orientation remains a Z-helix but the angle is much steeper than in the S1 cell wall layer (Figure 2B). The S3 layer is very thin and located next to the cell lumen. The microfibril orientation is shifted back clockwise from a Z-helix to a S-helix from the outer to the inner part of the layer (Figure 2B)[8-12].

3 WOOD DECAY IN ANOXIC ENVIRONMENTS

Wood is degraded by both biotic and abiotic mechanisms. Wood exposed in natural environments is decayed by a combination of mechanical and chemical weathering as interactions with wind, precipitation, UV light, heat, and hydrolysis reactions, oxidation reactions, and highly effective wood degrading fungi and insects[14-16]. The range of both abiotic and biotic breakdown mechanisms is considerably restricted for wood in waterlogged environments and thus the rate of decay slows down. The main mechanism for both abiotic and enzyme catalysed biotic degradation of carbohydrates is hydrolysis of acetal linkages in cellulose and hemicellulose. The carbohydrates are thereupon depolymerised and solubilised[15]. In anoxic waterlogged environments the abiotic acid hydrolysis of the carbohydrates is extremely slow due to the mild acid conditions, the low temperature, the low pressure in the submerged environment, and the very slow reaction rate of autohydrolysis. The half-life of cellulose in neutral pH is estimated to be about 100 million years[17]. The aromatic structure of lignin with delocalised electrons provides a range of possible redox and radical reaction which break down the ether, ester and carbon linkages of the lignin molecular structure.

Abiotic decay by slow hydrolysis was the predominant explanation for decay in relation to waterlogged archaeological wood in anoxic environments[18] until it was generally accepted in the 1990's that bacteria are responsible for the main decay[19-21]. Possible decay by wood degrading bacteria was suggested in the 1940's and 1960's with light microscopy studies[22-24] and verified by studies using scanning and transmission electron microscopy in the 1980's[25-26]. Even though biological decay factors are now agreed to be the most important, abiotic decay has been observed in a single study: Waterlogged archaeological hardwood and softwood 900-2200 years old without observable microbial decay pattern appeared to show a general weakening and delamination of the cell wall layers[27]. This observation has not yet been confirmed by other studies over the past 35 years. However, since bacterial decay has been recognised and has been the main focus of attention abiotic decay has not been studied in as much detail and more research is needed within this field.

3.1 The anoxic environment

Living organisms derive the necessary energy for cellular functions through oxidation of a suitable energy source. Heterotrophs derive this energy from organic substrates such as plants. At the end of the oxidative electron transport chain the electron is transported to a terminal electron acceptor. Oxygen is used as a terminal electron acceptor in aerobic respiration, but when the oxygen level drops drastically the redox potential of the soil/sediment is decreased as other terminal electron acceptors are used. The metabolic energy output is however not as high for the anaerobic reduction processes as for the aerobic respiration. The electron acceptor used is determined by the redox potential of the soil/sediment. The redox potential is a function of pH and each electron acceptor can be used in a quite broad range depending on the pH of the soil/sediment. The sequence of terminal electron acceptors for neutral soils/sediments is listed in Table 1.

In water saturated soils and sediments oxygen is only available by diffusion. Diffusion is slow and the oxygen concentration decreases with the depth of burial and concentration of organic particulate material. This leads to aerobic respiration in the uppermost layer of a waterlogged ecosystem and sulphate reduction and/or methanogenesis in the lower layers of the soil/sediment. Due to spatial variations in the system more than one electron acceptor is often used at a given time in a waterlogged ecosystem[28-30].

Table 1 *Sequence of terminal electron acceptors for neutral soils/sediments listed with the highest oxidation state first. Data from Killham[28] and Standing & Killham[30]*

Redox potential (pH 7)	Terminal electron acceptor	Final product	Associated microbial process
+ 820	O_2	H_2O	Aerobic respiration
+ 420	NO_3^-	N_2	Denitrification
+ 410	Mn^{4+}	Mn^{3+}	Manganese reduction
+ 400	Carbohydrate	Organic acid	Fermentation
- 180	Fe^{3+}	Fe^{2+}	Iron reduction
- 200	NO_3^-	NH_4^+	Dissimilatory nitrate reduction
- 220	SO_4^{2-}	H_2S	Sulphate reduction
- 240	CO_2	CH_4	Methanogenesis

3.2 Cellulolytic microorganisms

A great variety of cellulolytic bacteria and fungi are able to convert insoluble cellulosic substrates to soluble sugars that can be used as an energy source by the cell. The process is catalysed by a variety of different enzymes; the cellulases. These enzymes hydrolyse the β(1-4) glucosidic bonds in cellulose. Due to the intermolecular bonding in the crystalline structure multiple cellulase enzyme systems acting in synergy are needed to efficiently degrade cellulose. The energy and cell mass yield of aerobes per hydrolysed sugar unit is higher than that of anaerobes. This requires the anaerobic bacteria to evolve a highly effective system for extracellular degradation of polymeric substances. A strategy unique to bacteria is the organisation of cellulases in multi-component complexes known as cellulosomes attached to the bacterial cell wall[17,31-33]. Hemicelluloses are amorphous due to their branched and substituted structure and the carbohydrates are more readily accessible to enzymes than the glucose molecules in cellulose. Hemicelluloses are therefore easier to hydrolyse but a wide range of different hemicellulases are needed to de-branch and depolymerise the complex structure of hemicelluloses[34].

3.3 Lignin degradation

A wide range of anaerobic bacteria that are able to hydrolyse cellulose and hemicelluloses cannot degrade lignin. The general absence of oxygen prevents the known oxidative degradation, why lignin is fare more recalcitrant in anoxic environments. This also means that the carbohydrates will be less accessible for degradation as the lignin physically blocks the cellulose and hemicellulose. However, a few studies show slow but incomplete microbial decay of high molecular weight lignin in anoxic environments[35-37]. In addition lignin monomers, low-molecular weight oligomers, and chemically modified lignins have been shown to be significantly degraded by anoxic bacteria[36-42]. But the pathways and transformation mechanisms still remain uncertain[40].

4 MICROBIAL DECAY OF WATERLOGGED ARCHAEOLOGICAL WOOD

Bacterial decay patterns in wood differ from those produced by wood decaying fungi, just as the bacterial decay patterns also differ from each other. It is therefore possible to characterise different types of microbiological wood decay by the organisms and the unique decay pattern they produce at the cell wall level. In waterlogged ecosystems with low amounts of oxygen soft rot fungi and tunnelling bacteria dominate the degradation of wood. In near anoxic and possible also completely anoxic environments where decay by aerobe wood decaying organisms is completely suppressed the degradation of wood is dominated by erosion bacteria[43-46].

Microscopic studies of several different wood species of waterlogged wood show that erosion bacteria are the main degraders of wood in both terrestrial and aquatic waterlogged environments under low oxygen conditions. The decay rate is very slow but erosion bacteria are active within a great span of wood species and waterlogged ecosystems[19,21,45,47-50]. Arctic wood millions of years old has been observed to have the same decay pattern as foundation piles only a few hundred years old[1,19,50].

Despite the main decay by erosion bacteria it is not uncommon to find aerobic microbial decay patterns in archaeological wood in the surface layer of the objects[19,49,51,52]. The different decay types found in waterlogged archaeological wood tell an environmental history of the objects. If erosion bacteria decay is the only decay pattern the wood has most likely been waterlogged close to the felling date whereas presence of brown and/or white rot fungi tells that the object must have had some years of above ground service life before it became waterlogged[19].

5 IDENTIFICATION OF EROSION BACTERIA

Attempts have been made to isolate and identify wood degrading bacteria in pure cultures[53]. Interestingly it has so far not been possible to grow monocultures of true wood degrading bacteria acting in waterlogged environments or to reproduce the decay patterns with pure cultures in the laboratory[54]. The most promising attempts have been the isolation of single bacteria by the use of optical tweezers and micro-dissection combined with sub-culturing on pine wood, kapok fibres and cellophane as substrates. This has led to a severe reduction in the number of bacterial species, but not monocultures[55-57]. Comparison of DNA extracted from the sub-cultured bacterial isolates and DNA extracted directly from four different wood species from 19 European archaeological sites has led to the conclusion that erosion bacteria most likely fall within the *Cytophaga-Flavobacterium-Bacteroides* sub groups although the *Pseudomonas*, *Cellvibrio*, and *Brevundimonas* groups have also been commonly found in waterlogged wood[58]. Other attempts to identify erosion bacteria with molecular biological techniques have been performed in combination with traditional anaerobic culturing of bacteria extracted from waterlogged ash wood and by direct extraction from waterlogged wood samples. This has demonstrated that many bacteria from heavily decayed waterlogged wood belong to families which have lignocellulolytic capacity, but also that a major part of the bacteria found are unclassified and therefore unknown[59-60]. A general hypothesis is that a consortium of bacteria is responsible for the wood decay. This is supported by the facts that it has been impossible to isolate pure strains of erosion bacteria, that other types of bacteria (secondary degraders) have been observed by transmission electron microscopy in wood cells in relation to erosion bacteria decay, and that cross-feeding is a well known strategy in natural ecosystems[33,45,50,57,61,62].

6   MORPHOLOGICAL   DECAY   PATTERNS   IN   WATERLOGGED
     ARCHAEOLOGICAL WOOD

Attack by soft rot fungi, tunnelling bacteria and erosion bacteria starts from the surface of the wood and progresses inwards[63,64]. The decay is slow and a piece of wood is often heterogeneously decayed due to a gradient of decay from the surface to the core. The surface is soft and spongy while the core is often sound; only totally degraded material has a homogeneous soft structure from the surface to the core[4,65,66]. Zones with different degrees of degradation can be distinguished by differences in hardness, density, strength, porosity, water content, and often also colour. The basic density can be lower than 100 kg/m^3 for totally decayed wood and up to values close to sound wood for the same species (400-700 kg/m^3)[8,67]. Despite extensive decay the dimensions of the object and surface details such as tool marks and ornamentation remain unaltered in waterlogged conditions. This leaves the wood with a high archaeological value. Uncontrolled air drying will lead to irreversible alterations such as collapse, shrinkage, splitting and warping (Figure 3)[67]

Figure 3 *Section of a 12 m long oak dugout canoe from Varpelev, Denmark radiocarbon dated to approximately 1030 BC. The core of the log is still sound but the surface layer of the log is heavily decayed and has suffered severe damage due to air drying. The dugout is now under re-conservation at the National Museum of Denmark. Picture ©: Bevarings-center Øst, Denmark*

A

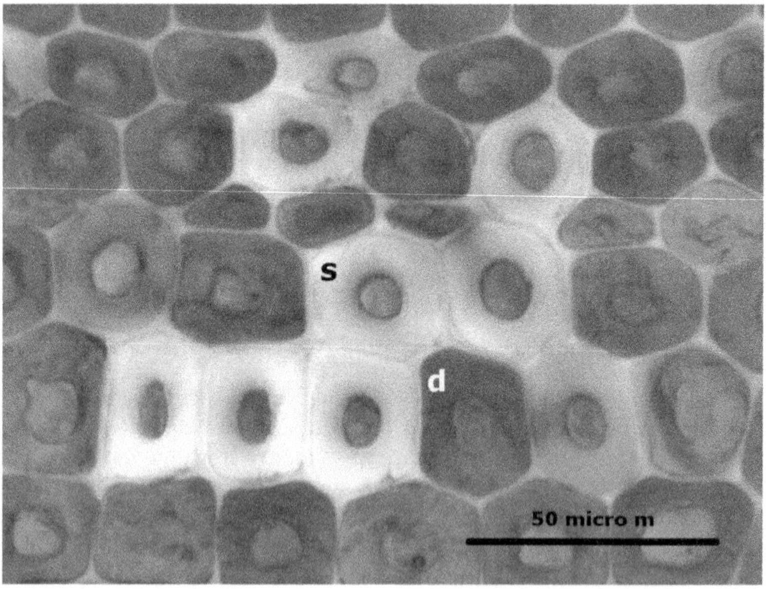

B

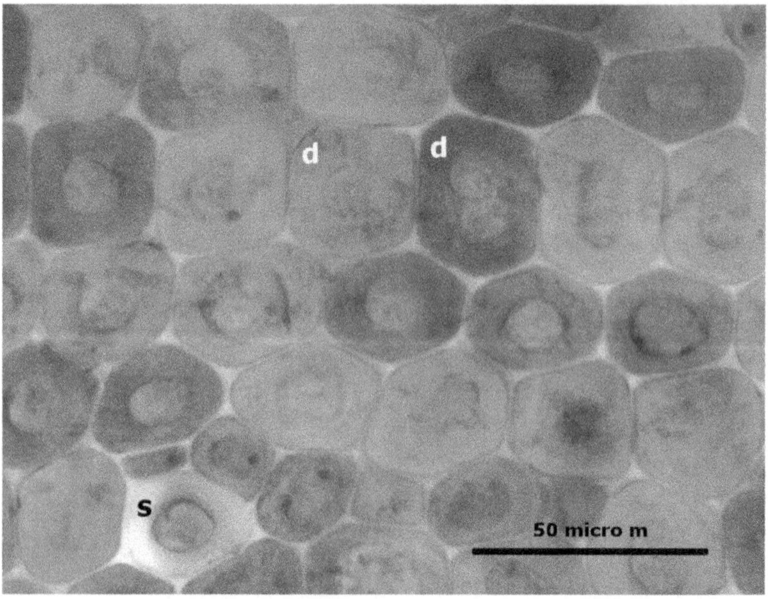

Figure 4 *Light microscope cross section of a Norway spruce (*Picea abies *(L.) Karst) pole decayed by erosion bacteria. The pole was placed in the moat surrounding Copenhagen (Denmark) in medieval times (1614 ± 15 AD). A: Medium decayed wood with a mix of sound (s) and decayed (d) tracheids. B: Heavily decayed wood with only a single sound tracheid. Pictures ©: Nanna Bjerregaard Pedersen*

due to changes in the physical and chemical properties of the wood material[68,69]. The degree of distortion varies from superficial to total disintegration depending on the degree of degradation[4,43].

6.1 Erosion bacteria decay at the cell level

The invasion of erosion bacteria starts from the wood surface, through rays and pits to the cell lumen. The decay leads to a complete destruction of the secondary cell wall. The bacteria leave a residual material that spreads and in some instances constricts or even fills up the whole cell lumen. The residual material does not have birefringence in polarised light. This indicates a breakdown of the crystalline cellulose microfibrils[45]. As the lignin-rich compound middle lamella is not degraded, the morphology of the wood is preserved and it is possible to distinguish different cell types even in heavily decayed wood. Cross sections of typical erosion decay pattern for moderately decayed wood is observed as a heterogeneous decay with sound cells adjacent to heavily degraded cells (Figure 4)[62].

In heavily decayed wood only a small proportion of sound cells are observed (Figure 4) whereas totally decayed wood only consists of a fragile skeleton of the compound middle lamella and the residual material (Figure 5). The degree of degradation is proportional to the number of decayed wood cells. By microscopic examination of a given point in the sample the degree of degradation can be classified by evaluating the relative proportion of sound and decayed cells[19,50,70].

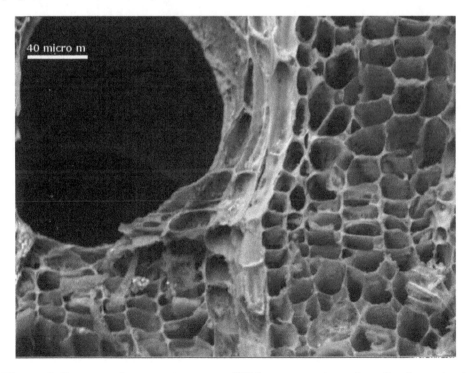

40 micro m

Figure 5 *Scanning electron microscopy (SEM) cross section of totally decayed ash* (Fraxinus Excelsior *L.) where only a fragile skeleton of the compound middle lamella is left. Sound ash wood is shown in Figure 2A. Due to freeze drying of the material the cells have not collapsed. Picture ©: National Museum of Denmark*

The heterogeneous decay of adjacent sound and decayed cells seen in cross section suggest that one wood cell at a time is attacked. However cross sections only provide information in one plane of the whole cell length. In longitudinal sections the erosion bacteria decay is seen as either a stripy appearance of the wood cell wall[26,70] or diamond- or V-shaped patterns of alternating sound and decayed cell wall (Figure 6)[19,22,26,50,63].

The differences between the stripy and diamond-shaped decay patterns have been suggested to depend on oxygen limiting conditions resulting in a more incomplete utilisation of the wood substrate[45]. However, other explanations such as different species of erosion bacteria acting on the material and differences in density and chemical composition of the cell wall material are just as likely. Longitudinal sections viewed in polarised light show alternating birefringence and non-birefringence areas i.e. areas with crystalline and amorphous structures[26,63]. Examination of a series of 2D cross sections along 340 μm of the same cluster of tracheids in the longitudinal direction showed that an apparently sound tracheid is not necessarily sound in the whole length of the cell. In addition it was observed that a moderately degraded tracheid consists of both degraded and sound areas[71]. This shows that unless total degradation is present tracheids are heterogeneously decayed within the same cell.

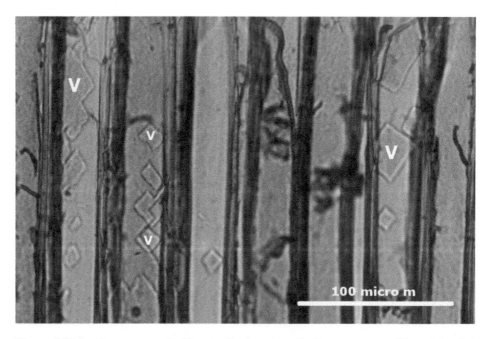

Figure 6 *Light microscope radial longitudinal section of a Norway spruce (*Picea abies *(L.) Karst) pole decayed by erosion bacteria. The pole was placed in a moat surrounding Copenhagen (Denmark) in medieval times (1614 ± 15 A.D). A diamond or V-shaped pattern (V) of alternating sound and decayed cell wall material is seen. Picture ©: Nanna Bjerregaard Pedersen.*

Differences in decay patterns are also observed at the cell type level. Ray tracheids and ray parenchyma are generally well preserved in pine and spruce compared to axial tracheids[63]. Differences in recalcitrance of different cell types in waterlogged archaeological wood were also reported in 1983[72]. Differences in decay were found to correlate to changes in lignin composition and lignin concentration but not to the physical accessibility to the actual cell wall. Vessel elements in hardwood with predominantly guaiacyl lignin had a low degree of decay whereas the secondary wall of fibre cells with high content of the less condensed syringyl lignin had a high degree of deterioration[72].

Physical accessibility is not the primary factor of decay at the cell wall level but it seems to be important for the possibility of the bacteria to reach the individual cells. In vessel elements and in large and simple cross-field pits in pine the initial decay is in close association with these easy accessible transport routes. In spruce and fir with small bordered cross-field pits the initial decay is not necessarily observed close to rays[50,63].

6.2 Erosion bacteria decay at the ultrastructural level

Observations at the ultrastructural level by electron microscopy show erosion bacteria decay as formation of narrow erosion troughs in the cell wall. Erosion bacteria are short rod-shaped Gram negative bacteria 3-5 μm long. Groups of bacteria are aligned close to each other forming individual decay troughs producing a ribbed cell wall surface (Figure 7). The attack on the cell wall is initiated from the cell lumen and progress towards the compound middle lamella. Bacteria attach themselves to the S3 cell wall layer by means of a biofilm with their main axis orientated parallel to the cellulose microfibrils and in close intimate association with the surface of the tracheid wall. The S3 layer is penetrated locally and the attack progresses into the S2 cell wall layer. The bacteria are firmly attached to the cell wall and move by gliding. The contact zone between bacteria and cell wall consists of a loosely textured wall material and membrane-bound vesicles. This suggests the involvement of extracellular enzymes complexed in cellulosomes.

Where several bacteria are closely packed on the cell wall the individual erosion troughs merge and the whole S2 cell wall is gradually removed (Figure 8)[26,45,57,63,73,74]. The S1 and S3 cell wall is only degraded in advanced stages of decay (Figure 8). The S3 cell wall can often be seen overlaying the residual material. This has been explained by the higher lignin content of the S3 cell wall compared to the S2 cell wall[44,45,63]. The lignin concentration is not remarkably higher in the S1 and S3 cell wall compared to the S2 cell wall[12]. Chemical composition and cellulose micro-fibril angle and direction might also influence the more recalcitrant nature of the S1 and S3 cell wall layers. The compound middle lamella is most often reported as unaffected even at very advanced stages of decay. But reduction in intensity of staining, loss of electron density and occasional degradation all suggest minor modification of lignin in the middle lamella[44].

The residual material is believed to consist of residual products from cell wall decay as lignin or lignin degradation products mixed with bacterial slime and secondary degraders. But the chemical composition of the material is not known[26,45,49]. Secondary degraders are observed to utilise the residual material for their nutrition thereby changing the density of the residual material. This is supported by presence of irregular clearing zones surrounding single or colonies of secondary degraders and by a change in the composition of the residual material surrounding these bacteria[45,63].

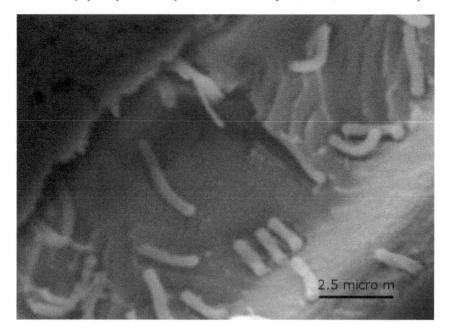

Figure 7 *Scanning electron microscopy (SEM) micrograph of erosion bacteria. The bacteria form individual troughs in the cell wall and thereby produce a ribbed cell wall surface. Picture ©: Charlotte Gjelstrup Björdal*

6.3 Tunnelling bacteria

The morphology of tunnelling bacteria decay differs markedly from erosion bacteria decay. Tunnelling bacteria also settle on the S3 cell wall on the lumen side with the aid of a biofilm. The bacteria penetrate into the S2 cell wall were they produce tunnels in all directions. The different cell wall layers are not resistant to decay and the lignin rich middle lamella is also penetrated and degraded. Only one bacterium is present in each tunnel, but cell division leads to branching of the tunnels. The tunnels never cross each other but in severely degraded wood the tunnels collapse and the whole cell wall is degraded. Cross walls or tunnel bands are formed in the tunnels and are believed to contain extracellular slime used for moving. Otherwise the tunnels are not filled with residual material suggesting that the bacteria are able to utilise all cell wall components including lignin. Tunnelling bacteria can tolerate a wide range of pH and temperatures and have the ability to degrade both softwood and hardwood but also preservative treated timbers and durable timbers with high lignin and/or extractive content[21,26,75].

6.4 Cavitation bacteria

Cavitation bacteria have been suggested as a third type of wood degrading bacteria, but is rarely reported. It has been suggested that it might be a form of erosion bacteria found in restricted situations[44]. Cavitation bacteria also attach to the surface of the wood cell wall by use of extracellular slime. The bacteria penetrate the S3 cell wall by forming a small hole and produce angular or diamond shaped cavities in the S2 and S1 cell wall but not in the middle lamellae. Unlike tunnelling and erosion bacteria decay the degradation is not localised on the cell wall and it has been suggested that the bacteria does not immobilise its enzymes on cellulosomes but rather produce diffusible enzymes which can penetrate the wood cell wall [26,74].

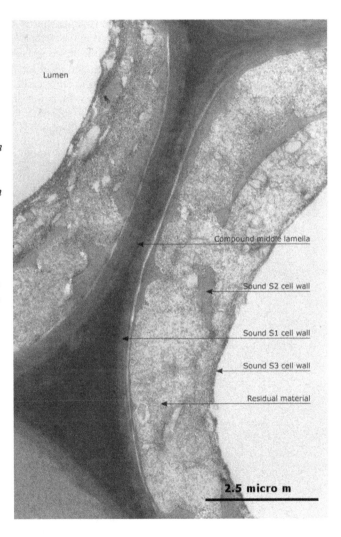

Figure 8 *Transmission electron microscopy (TEM) of ultra-thin cross sections of early wood from a Norwegian spruce (*Picea abies *(L.) Karst) pole decayed by erosion bacteria. The pole (dated to 1614 ±15 AD) has been excavated from the moat that surrounded Copenhagen (Denmark) in Medieval times. The cross section shows details of the cell wall of two adjacent tracheids with intact cell corners, compound middle lamella, S1 and S3 cell wall layers. The S2 cell wall in both cells is decayed. A residual material is left behind, but minor S2 cell wall areas are still intact. TEM cross section of sound spruce tracheids is shown in Figure 2C. Picture ©: Nanna Bjerregaard Pedersen*

6.5 Soft rot

Soft rot decay is caused by the two taxonomically subdivisions *Ascomycota* and *Fungi imperfecti*. The fungi grow in the S2 cell wall where the hyphae produce either cavities in the cell wall or erode the cell wall. A large number of different species produce round or oval cavities (Figure 9). The cavity micro-morphology varies depending on the fungus, the substrate, and the growth conditions. Colonization of wood is first observed in rays and vessels from where hyphae grow into the fibre/tracheid lumen. The hyphae bore into the secondary wall through a small borehole in the S3 cell wall and align their direction of growth parallel to the direction of the cellulose microfibrils. Cavity formation takes place in the S2 cell wall but has also been observed in the S1 layer. In advanced stages of decay individual cavities enlarge and coalesce until the entire S2 cell wall is depleted. The middle lamella is normally not degraded, only penetrated. Soft rot fungi attack tracheids, fibres, axial and radial parenchyma cells, ray tracheids, and vessel elements. The carbohydrates are decayed in preference to lignin, but some modification of lignin by soft rot fungi does occur. The type and quantity of lignin in the substrate influences the rate of decay. Hardwoods are more susceptible to decay than softwood[38,76-78].

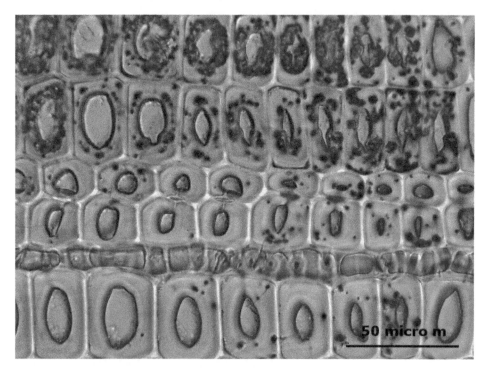

Figure 9 *Light microscope cross section of softwood decayed by soft rot. The bore holes from the hyphae in the secondary wall are easily distinguished. The decay pattern is clearly different from erosion bacteria decay (see also Figure 4). Picture ©: Charlotte Gjelstrup Björdal*

7   CHEMICAL COMPOSITION OF WATERLOGGED ARCHAEOLOGICAL WOOD

Knowledge of the chemical composition of waterlogged archaeological wood from a relatively wide range of publications is unfortunately obscured by the fact that with the exception of a few cases the decay pattern(s) and decay organisms of the wood have not been established before analysis[1,77,79-81]. The wood is most often solely defined by wood species and as waterlogged and/or archaeological but not in terms of the morphological decay type. Since the different decay organisms facilitate different ultrastructural patterns it must be expected that the chemical composition is different in wood decayed by different microorganisms. In addition the picture is complicated by the fact that slow abiotic decay can have some influence on the chemical composition in especially weak and moderately decayed wood. A greater understanding of the microbial degradation in relation to the chemical fingerprint left in the wood will also result in a better ability to distinguish between microbial degradation and abiotic degradation of wood in anoxic environments.

As waterlogged archaeological wood is a heterogeneous material with several factors influencing the chemical composition, it is hard to identify general patterns of the chemical composition of the wood decay. The most important are decay type, submerging environment, degree of decay, wood species, and the state of the sound wood. Some of these factors can be difficult if not impossible to establish for the whole post depositional history of a given sample.

7.1 Analytical methods

The most widely used methods for chemical characterisation of waterlogged wood are traditional wet chemical methods to separate and determine quantitatively the chemical constituents of wood[68,77,79,82-87], Fourier Transform Infra-Red spectroscopy (FTIR)[79,80,85,88-92,114], Nuclear Magnetic Resonance spectroscopy (NMR)[1,86,93-100], and analytical pyrolysis coupled to Gas Chromatographic and/or Mass spectroscopy (Py-GC/MS)[1,86,99-104].

An overall characteristic feature of the methods is that they all provide an average picture of the chemical composition of the material studied. They are all based on milled wood or otherwise undefined average areas of the material. The analyses give results on the average composition of the material. But unless the material is morphologically sound or totally degraded and only the compound middle lamella and the residual material is left the material consists of a mixture of sound and degraded S2 cell wall areas, intact compound middle lamella, and partly decayed S1 and S3 cell wall layers. The observed chemical composition will thus arise from an average of the residual material left after the secondary wall, chemical changes in the morphological intact compound middle lamella, the S1 and S3 cell wall, and from the sound cell wall areas present in the material. This average composition will not establish how the remaining cell wall areas and the decayed cell wall areas are composed. Combination of high resolution imaging and spectroscopic techniques can be done with chemical imaging[105]. These techniques can give spatially resolved information on the molecular structure with a resolution as low as 0.25 μm[106,107]. This makes it possible to distinguish not only between different cell types and decay stages but also on the individual cell wall layers. This has so far been carried out on waterlogged archaeological wood with UV-micro-spectrophotometry and Raman imaging[108,109].

Despite the general lack of determination of decay type all chemical compositional studies show the general trend of a preferential loss in carbohydrates compared to the lignin fraction of the wood. Microscopic studies, determination of density, or the maximal water content combined with chemical analysis have shown that greater loss of carbohydrate and relative increase in lignin concentration correlates with degree of degradation. The higher the degree of decay the more lignin and the less carbohydrate are present relatively in the material[51,79,100]. This correlates well with the micro-morphological features of preferential decay of the cellulose rich secondary wall, the loss of birefringence in the decayed wood cell wall and the intact lignin rich middle lamella mainly in erosion bacteria degraded wood[110]. Several of the methods are therefore useful for determining the degree of degradation. In particular FTIR analysis and Direct Exposure Mass Spectroscopy (DE-MS) are rapid, non-destructive, and accurate methods for measuring the extent of degradation[90,101,111-114].

7.2 Water soluble carbohydrates

The amount of water soluble carbohydrates determined with hot and/or cold water extraction is significantly smaller in waterlogged archaeological wood of both hard- and softwood compared to sound reference samples[79,84,86]. The changes though have been shown not to be associated with the degree of erosion bacterial attack[79]. The water soluble carbohydrates are either lost due to diffusion during the long storage in waterlogged conditions or due to consumption by microorganisms. It has generally been assumed that plant material is first colonized by bacteria that feed on the easy accessible sugars and later the plant material is colonized by bacteria (and fungi) that are able to degrade the lignocellulosic cell wall[21,33]. The results also indicate that the bacteria eventually in combination with secondary degraders utilize the carbohydrates liberated from the cell wall quite effectively.

7.3 Structural carbohydrates

The structural carbohydrates are preferentially lost in proportion to lignin in microbial decayed waterlogged wood, but even in heavily decayed wood submerged for thousands or millions of years some carbohydrates are preserved[1,97]. This is due to differences in the cell wall composition of the individual cell wall layers. The lignin and carbohydrate composition and concentration are important factors as well as the three dimensional construction of the individual cell wall layers. Unaltered compound middle lamellas, S1 cell walls, and S3 cell walls in wood decayed by erosion bacteria will leave small amount of carbohydrates in the material even in heavily decayed wood without any sound S2 cell wall layers left.

Analyses of moderately decayed waterlogged oak with X-ray scattering diffractometry have shown that the crystallinity of the intact cellulose microfibrils does not change compared to sound reference wood, but the degree of polymerisation is reduced for the waterlogged samples[92,114]. Crystallinity determinations with X-ray diffraction of plant material are not straight forward[115], but the results are supported by determination of α-, β-, and γ-cellulose with traditional wet chemical methods. In samples with weak, moderate and severe erosion bacterial decay the carbohydrate content is less than in sound reference material, but the degree of polymerization of the cellulose does not seem to change much compared to the sound waterlogged wood from the same sample material[79]. High resolution solid-state NMR shows a decrease in peaks assigned to carbohydrates but no line broadening of the signals and no appearance of new signals. This is a strong indication that chemical rearrangement inside the biopolymer network does not occur[93,94]. The above mentioned studies indicate that erosion bacteria utilise the cellulose effectively. This is in agreement with the ultrastructural observations on erosion bacteria decay. Loosely attached or partly crystalline un-wound microfibrils are not observed. A cell wall area is either sound or it is converted to an amorphous residual material (Figure 8).

The hemicellulose fraction has, as expected, been reported to be more susceptible to decay than the cellulose[1,81,85,86,93,100]. This is due to both the greater resistance of the crystalline cellulose compared to the amorphous hemicellulose and also to ultra-structural hindrance.

7.4 Lignin

It has generally been shown that the chemical structure of lignin is conserved to a high degree; only minor alternations of the lignin structure have been reported. The difference in the reported quantitative and qualitative structural changes of lignin in waterlogged wood is probably due to the fact that the decay patterns and the degree of degradation of the material haven't been established in most cases, that the submerging environment are different, and that the applied analytical methods have different sensibility.

7.4.1 Quantitative lignin studies. Quantitative studies on the mass loss of lignin have only been conducted in a few cases. A loss of 90-98 % of the carbohydrate fraction and about 15-25% of the lignin fraction have been estimated in a study done on 2,000 and 25,000 years old hardwood (alder and oak) with wet chemical analysis, ^{13}C solid state NMR, Py-GC-MS, and density measurements. The description of the micro-morphology of the material based on observations with scanning electron microscopy has been reported as a typical erosion bacteria decay pattern[81]. Sandak and coworkers[14] found no significant evidence for a decrease in the lignin content in waterlogged oak based on traditional wet chemical methods combined with density measurements. The samples had a notable carbohydrate loss but were not as heavily decayed as the alder and oak wood described by Hedges et al.[81]. Gel Permeation Chromatography (GPC) analyses of the molecular weight distribution of lignin have shown a slight decrease in molecular weight and thereby a minor depolymerisation of the lignin[96,99].

7.4.2. Qualitative lignin studies. Qualitative studies on the chemical changes in the lignin structure have been applied more widely. ^{13}C solid-state NMR studies on waterlogged wood show little chemical change in the lignin composition compared to sound reference wood. The most frequent inter-monomeric bond in lignin, namely β-O-4, is preserved[93,97,100]. More detailed 2D ^{13}C, 2D-HSQC, and ^{31}P NMR studies on hardwood and softwood lignin extracted from waterlogged wood have shown that the principle intermonomeric bonds in lignin (β-O-4, β-5, and β-β) had a high degree of conservation. No significant changes were observed in the relative amounts of the β-β bonds compared to sound reference wood. A slight decrease in the amount of β-O-4 units was observed together with a significant reduction of the β-5 intermonomeric bonds in hardwood lignins. The 5-5'-O-4 intermonomeric bond could not be detected in waterlogged wood[95,96].

Combination of wet chemical analysis, ^{13}C solid state NMR, and Py-GC-MS on alder and oak wood with a characteristic pattern of erosion bacteria decay showed in general good preservation of the lignin structure but a selective loss of syringyl units over guaiacyl units[81,103]. Decrease in syringyl units of hardwood samples has also been reported by Giachi & Pizzo[80], Iiyama et al.[84], and Richards and West[102]. Lucejko and colleagues[101,116] have in contrast found the syringyl/guaiacyl ratio in archaeological hardwood lignins (beech and elm) to be comparable to sound reference wood with Py-GC-MS analysis. This was also seen by Pan and colleagues[86] on isolated hardwood lignin from well preserved waterlogged hardwood from a river site. Hardwood lignin of the secondary wall is relatively rich in syringyl units. A selective loss of syringyl units corresponds well with the morphological characteristic of the preferential decay of the secondary wall for both erosion bacteria and soft rot fungi[84], and the general observation that guaiacyl lignin is more stable than syringyl lignin[117].

Demethylation of the lignin moieties in waterlogged wood have been shown in several cases. 1.1-56 million years old Canadian arctic softwoods with a morphology described as a typical erosion bacteria decay have shown that the remaining lignin have lower levels of methyl groups on lignin units than lignin from a sound reference sample[1]. Demethylation of lignin from waterlogged wood has also been reported by Colombini et al.[96], Lucejko[101], Lucejko et al.[116], and van Bergen et al.[104]

The amount of oxidized phenols (e.g. vanillic and syringic acid) in the lignin moiety of waterlogged wood does not show a general trend in the studied sample material. This indicates that the amount of oxidized phenols to a great extent depend on the decay type and/or burial environment. A combined study with wet chemical analysis, ^{13}C solid state NMR, and Py-GC-MS on alder and oak wood with a characteristic pattern of erosion bacteria decay did not show significant amount of oxidized phenols[76,103]. This tendency was also reported in archaeological hardwood from a fresh and marine water site[104] and softwood from a lagoon[95]. The low levels of oxidized lignin units compared to sound reference samples indicate anaerobic decay of the lignin in these samples. Several studies on waterlogged soft- and hardwood from the ancient harbour of Pisa show elevated amounts of oxidized lignin units and enhanced amounts of carboxylic acid functionalities[80,99,96,101,111,113,116,118]. This suggests oxidative lignin degradation. The excavation site is in several cases reported as anoxic but the morphological decay pattern has only been reported in one case as a combination of soft rot and erosion bacteria decay[88]. The oxidative lignin decay may have been facilitated by erosion bacteria since they are believed to be facultative anaerobic bacteria and might be able to decay lignin both aerobic and anaerobically. It is also possible that soft rot is the only microbe responsible for the oxidative decay since it is an obligate aerobic fungi. Other obligate aerobic fungi as brown and white rot might also have degraded the wood before it was embedded in the anoxic sediment. Giachi et al.[118] suggest degradation by white rot fungi in sample material from the ancient harbour of Pisa after chemical analysis of the samples. An explanation could also be that the reported oxidative lignin degradation is due to still unknown mechanisms of anaerobic lignin decomposition.

Side chain degradation of lignin in waterlogged wood has been reported in a few cases. Obst et al.[1] have found side chain degradation in 1.1-56 million years old Canadian arctic softwoods with a micro-morphology similar to erosion bacteria decay pattern. Lucejko[101] and Lucejko et al.[116] found a lower amount of intact propyl side-chains indicating lignin side-chain degradation in marine timbers from the San Rossore Roman Harbour (Pisa).

Studies on lignin extracted from waterlogged wood samples from the San Rossore Roman Harbour (Pisa) also show that lignin has undergone partial depolymerisation with cleavage of ether inter-monomeric bonds leading to production of a higher concentration of free phenol groups on both guaiacyl and syringyl units. The increased phenol content was mainly due to condensed phenols[96,111]. Salanti and colleagues[99] did not however find notable elevated amounts of free phenols in oak and strawberry tree (*Arbutus unedo*) from the San Rossore Roman Habor (Pisa). Pan et al.[86] found a higher phenolic content in isolated lignins from well preserved waterlogged hardwood from a river site with a higher degree of condensation. A higher degree of condensation in the remaining lignin in waterlogged samples compared to sound reference samples have also been reported in 1.1-

56 million year old Canadian arctic softwoods with a micro-morphology similar to an erosion bacterial decay pattern[1].

7.5 Organic extractives

Examination of the content of organic components extracted with organic solvents and/or water in waterlogged archaeological wood and foundation piles have shown a reduced content compared to sound wood[79,85,86,114]. The changes cannot be correlated to the degree of bacterial attack[79]. The major constituents of the organic extractives measured with GC and GC-MS analysis were qualitatively similar but quantitatively different[86]. This indicates that water flow on the burial site and not microbial degradation is the major cause of depletion of organic extractives. This is supported by a study showing that oak wood from a lake and a river sites had a lower extractive content than oak samples from a peat bog and from the soil[114].

7.6 Inorganic components

A general trend of waterlogged archaeological wood is that the inorganic content is increased compared to sound reference samples[1,77,82,85-87,114,118]. The elevated amount cannot be directly associated with the degree of bacterial attack[79]. The inorganic level is merely caused by a combination of the permeability of the wood structure (wood species, degree of degradation), the level of inorganic components in the submerging environment, the water flow on the burial site, and contamination of cracks and fissures during burial. In cases where metals are found in close contact with waterlogged wood the inorganic levels can be many times greater than the level in sound reference samples[85]. The corrosion products will diffuse into the wood matrix and provide some physical stability of the wooden ultrastructure. The metal ions will be able to catalyze chemical degradation or inhibit biological degradation. The extent of chemical degradation will depend largely on the nature of the corrosion products and the submerging environment[102]. Marine timbers will often be associated with high levels of iron and sulphur due to corrosion of iron fastenings and release of sulphide in the anoxic environment[85,118,119]. Analysis of the inorganic content of both archaeological samples and foundation piles without any direct metal interference show increased nitrogen and phosphorus content with higher degree of degradation. Levels of sulphur, sodium, calcium, potassium, and iron do however not correlate with the degree of degradation[79,81].

8   ENVIRONMENTAL PARAMETERS CONTROLLING EROSION BACTERIA DECAY

Age is not the primary decay factor for waterlogged wood[48,69,80]. The waterlogged environment has a greater influence on the decay rates than the submerging age. It is however not finally established which factors are most important in relation to erosion bacterial decay. Oxygen concentration, redox potential, wood species, ion species concentration, salinity, pH, wave action/water flow, currents and sediment abrasion all seem to play a role[76,120].

It has not been fully established if erosion bacteria can thrive under completely anoxic conditions. This is hard to establish with *in situ* samples. Completely anoxic submerging environments may not have been anoxic in the entire submerging history of the artefact[120]. Laboratory experiments suggest that erosion bacteria are facultative anaerobes and even tolerate presence of hydrogen sulphide[56,70]. Lack of hydrogen sulphide in purified cultures in sub-culturing experiments suggests that sulphate-reducing bacteria are not directly related to degradation of lignocellulosic material[55]. Laboratory experiments with sound wood exposed to erosion bacteria and observations on waterlogged sound and archaeological wood salvaged from waterlogged environments suggest that erosion bacteria decay decreases with dept of burial. This is explained by higher growth rates with higher amounts of available oxygen[62,70,122]. On the other hand observations on 27 foundation piles from spruce, fir, pine, poplar, and oak solely decayed by erosion bacteria did not show decreasing decay with depth of burial[50].

It has been suggested that bacterial activity is related to water flow in the submerging environment. Observations on a high number of foundation piles showed that all wood species with an open structure had a higher water flow and were more susceptible to decay by erosion bacteria than less permeable wood structures[50,121]. Laboratory experiments with circulating water also showed a more intense decay compared to un-circulated replicas[70]. This correlates well with the observations that sapwood is less resistant to erosion bacteria decay than heartwood, softwood is generally more resistant than hardwood, and spruce more resistant than pine[19,49,50,80,122,123].

The pH tolerance for erosion bacteria is not known, but judging from their occurrence in many different environments they have a very broad pH tolerance[48,124]. The influence of ion species concentration on decay rates is not known either. Addition of nitrogen and phosphate to sediments in microcosm experiments did not promote bacterial wood decay[70]. Culturing of erosion bacteria in the laboratory showed that elevated levels of nitrogen, phosphate, sulphate, and cellobiose generally reduced or even prevented attack[56]. These observations are supported by a correlation between decreased nitrogen concentration in the soil and increased decay on wooden foundation piles. The same correlation was however not found for phosphate[48,125].

Environmental parameters controlling microbial decay in the waterlogged environments are important to establish and relate to the decay pattern, decay rate and chemical composition of the material. This is a very complex task but will be highly valuable in relation to a more in-depth understanding of the geochemical cycles of carbon, oxygen, nitrogen, and sulphur. A deeper insight into the parameters controlling erosion bacteria decay in anoxic environments is also directly applicable for historical wooden objects. Reburial and *in situ* preservation is becoming a gradually more used option when managing preservation of especially large shipwrecks or other large prehistoric constructions. The advantage is that the high cost of excavation, conserving and curating the finds are eliminated and the finds are saved *in situ* for future generations. But reburial and *in situ* preservation is not without cost; the sites have to be monitored continuously to secure the stability of the site[126,127]. A better understanding of the factors acting on waterlogged archaeological wood will lead to more qualified decisions when a preservation site is constructed and monitored.

9 SUMMARY AND PERSPECTIVES

In summary, wood is the largest input of biogenic carbon into the global carbon cycle. Under aerobic conditions, it is returned within decades as inorganic carbon mainly due to the action of wood decaying fungi. However anoxic ecosystems offer extreme environmental conditions and under these conditions the wood can be preserved for thousands of years. The recalcitrant nature of wood is not only controlled by the anoxic environment but also by the structure and organisation of the cell wall components. The decay mechanisms acting in anoxic ecosystems are though not fully understood. Preserved archaeological wood objects are due to their age, archaeological, and cultural record well suited for studying the degradation of biogenic carbon in anoxic environments. This provides both archaeological knowledge and knowledge of the chemistry and biology of the decay mechanisms and their interactions with the biopolymers in wood. Erosion bacteria are the main degraders of waterlogged archaeological wood but it has so far not been possible to isolate or explicitly identify the bacteria species. The decay starts at the surface and proceeds inwards leaving the wood with zones of different degree of decay. In cross section the typical erosion bacteria decay pattern is a mixture of sound and decayed cells adjacent to each other. In the longitudinal direction it is evident that the individual wood cells consist of a mixture of sound and decayed cell wall areas. In decayed cell wall areas the cellulose rich secondary cell wall is decayed and an amorphous residual material is left behind. The lignin rich compound middle lamella and often also parts of the S1 and S3 cell wall is preserved. This ultrastructural decay pattern corresponds well with the general observation that cellulose and hemicellulose is lost in preference to lignin. Hemicellulose is decayed in preference to cellulose. The degree of polymerisation is reduced for the remaining carbohydrates in waterlogged wood. Decayed cell wall areas have lost birefringence due to loss of crystalline cellulose, but the crystallinity of the remaining cellulose microfibrils is not significantly changed compared to sound reference wood. The lignin degradation is minor and incomplete. The chemical composition of the lignin is very similar to sound reference wood. The principle intermonomeric bonds have a high degree of preservation; and only minor changes due to de-methylation, side chain degradation, and slight depolymerisation have been observed.

The degradation of wood in waterlogged environments is dynamic and complex. The chemistry involved includes acid- and enzymatic hydrolysis of carbohydrates as well as a number of biotic- and abiotic redox reactions of which the latter is only rudimentary known. Furthermore the bacterial communities that have a biology and complexity of its own control the rate and extension of the decay. Nevertheless since billions of tons of wood are part of the global carbon cycle a better understanding of anoxic wood degradation will have a high value. There is a need for a deeper understanding of how the structural recalcitrance of wood, environmental factors, chemistry and microbiology interact.

Identification of erosion bacteria by molecular techniques and the ability to culture pure strains or purified consortia of erosion bacteria in the laboratory is needed to get a deeper understanding of the microbiology of anoxic wood decay. This would open a wide range of research opportunities to gain more solid knowledge on environmental requirements in relation to decay patterns and decay rates. By combining pure strain/purified consortia of erosion bacteria with sound wood it might be possible to establish if different species of erosion bacteria act in different wood species and waterlogged environments or if differences in the residual structure is related to degree of decay, environmental parameters as for instance oxygen concentration in the submerging environment, or solely to the lignin distribution, concentration and chemical composition in the cell wall material.

The chemistry of wood degradation in anoxic environments is not fully understood. A weak spot in our knowledge is the mechanisms by which lignin is degraded or conserved. This tie very closely to the microbiology involved and much can be learned by a parallel study of these two areas. This requires morphological characterisation of the degradation pattern prior to chemical analysis and/or combination of high resolution imaging and spectroscopic techniques to be able to distinguish different cell types, degradation stages, and cell wall layers.

The challenge in conservation of waterlogged archaeological wood remains is objects with a heterogeneous degree of degradation from surface to core. The sound core will to a great extend behave as sound wood in terms of chemical and physical properties. The totally decayed surface layer will act very differently from sound wood since the secondary wall material, which accounts for 70-90 % of the cell wall thickness, is lost. In addition the wood material situated between the totally decayed surface layer and the sound core will consist of a wood tissue with a mixture of sound and decayed cell wall areas. Exact knowledge on the residual structures on the molecular level will be highly important in order to optimise and develop conservation methods with the use of chemical impregnation agents.

None of these tasks are straightforward and requires a combination of research in several areas. But much can be learnt not only about basic chemistry and microbiology, but also on how we can use and interact with the largest component of the ecosystem – wood!

Acknowledgement

The authors would like to thank David Gregory from the National Museum of Denmark for professional support and proof reading and Lisbeth Garbrecht Thygesen from the University of Copenhagen for professional support. We also like to thank the National Museum of Denmark, the Viking Ship Museum, and Bevaringscenter Øst for permission to use pictorial material.

References

1   J.R. Obst, N.J. McMillan, R.A. Blanchette, D.J. Christensen, O. Faix, J.S. Han, T.A. Kuster, L.L. Landucci, R.H. Newman, R. C. Pettersen, V.H. Schwandt and M.F. Wesolowski, in *Tertiary Fossil Forests of the Geodetic Hills, Axel Heiberg Island, Arctic Archipelago,* vol. Bulletin 403, ed. R. L. Christie and N. J. McMillan, Energy, Mines, and Resources Canada, 1991, p. 123.

2   H. Thieme and R. Maier, *Archäologische Ausgrabungen im Braunkohlentagebau Schöningen*, Verlag Hahnsche Buchhandlung, Hannover, 1995.

3   M.E. Himmel and S.K. Picataggio, in *Biomass Recalcitrance. Deconstructing the Plant Cell Wall for Bioenergy*, ed. M. E. Himmel, Blackwell Publishing, Oxford, 2008, p. 1.

4   B.B. Christensen, *The Conservation of Waterlogged Wood in the National Museum of Denmark. With a report on the methods chosen for the stabilization of the timbers of the Viking ships from Roskilde Fjord, and a report on experiments carried out in order to improve these methods,* The National Museum of Denmark, Copenhagen, 1970.

5   P. Jensen, I. Bojesen-Koefoed, I. Meyer and K. Strætkvern, in *Proceedings of the 5th ICOM Group on Wet Organic Archaeological Materials, Portland/Maine 1993*, ed. P. Hoffmann, Ditzen Druck und Verlags-GmbH, Bremerhaven, 1994, p. 523.

6   I. Bojesen-Koefoed, in: *Proceedings of the 11th ICOM-CC Group on Wet Organic Archaeological Materials Conference, Greenville, North Carolina 2010,* ed. K. Strætkvern and E. Williams, The International Council of Museums, Committee for Conservation, Waterlogged Wood Working Group, 2012, p. 497.

7   S. Braovac and H. Kutzke, *J. Cultural Heritage*, 2012 (in press). http://dx.doi.org/10.1016/j.culher.2012.02.002.

8   J.L. Bowyer, R. Shmulsky and J.G. Haygreen, *Forest Products and Wood Science. An Introduction*, Blackwell Publishing, Oxford, 2001
9   D. Fengel, and G. Wegener, *Wood. Chemistry, Ultrastructure, Reactions*, Kessel Verlag, Remangen, 2003.
10 M. Fujita and H. Harada, in *Wood and Cellulosic Chemistry*, 2nd edition, ed. D. N. S. Hon and N. Shiraishi, Marcel Dekker, Inc., New York, 2001, p. 1.
11 P.J. Harris and B.A. Stone, in *Biomass Recalcitrance. Deconstructing the Plant Cell Wall for Bioenergy*, ed. M. E. Himmel, Blackwell Publishing, Oxford, 2008, p. 61.
12 S. Saka, in *Wood and Cellulosic Chemistry*, 2nd edition, ed. D. N. S. Hon and N. Shiraishi, Marcel Dekker, Inc., New York, 2001, p. 51.
13 M.A.T. Hansen, J.B. Kristensen, C. Felby and H. Jørgensen, *Bioresource Technol.*, 2011, **102**, 2804.
14 W.C. Feist and D.N.S. Hon, in *The Chemistry of Solid Wood*, ed. R. Rowell, American Chemical Society, Washington D.C., 1984, p. 401.
15 I.S. Goldstein, in *The Chemistry of Solid Wood*, ed. R. Rowell, American Chemical Society, Washington D.C., 1984, p. 575.
16 T.K. Kirk and E.B. Cowling, in *The Chemistry of Solid Wood*, ed. R. Rowell, American Chemical Society, Washington D.C., 1984, p. 455.
17 D.B. Wilson, in *Biomass Recalcitrance. Deconstructing the Plant Cell Wall for Bioenergy*, ed. M. E. Himmel, Blackwell Publishing, Oxford, 2008, p. 374.
18 P. Hoffmann and M.A. Jones, in *Archaeological Wood*, ed. R.M. Rowell and R.J. Barbour, American Chemical Society, Washington D.C., 1990, p. 35.
19 C.G. Björdal, T. Nilsson, and G. Daniel, *Int. Biodeter. Biodegrad*, 1999, **43**, 63.
20 R.A. Blanchette, *Canadian J. Botany-Revue Canadienne de Botanique*, 1995, **73**, S999.
21 R.A. Blanchette, T. Nilsson, G. Daniel and A. Abad, in *Archaeological Wood*, ed. R.M. Rowell and R.J. Barbour, American Chemical Society, Washington, DC, 1990, p. 141-174.
22 J.B. Boutelje and A.F. Bravery, *J. Inst. Wood Sci.*, 1968, **20**, 47.
23 L. Harmsen and T.V. Nissen, *Nature*, 1965, **206**, 319.
24 J. Liese, in *Holzkonservierung*, ed. F. Mahlke, E. Troschel and J. Liese, Springer-Verlag, Berlin, 1950, p. 44.
25 T. Nilsson and G. Daniel, in: *Biotechnology in the Pulp and Paper Industry: The 3rd International Conference*, Stockholm, June 16-19 1986, Swedish Forest Products Research Laboratory, 1986, p. 54.
26 A.P. Singh, and A.J. Butcher, *J. Inst. Wood Sci.*, 1991, **12**, 143.
27 K. Borgin, N. Parameswaran and W. Liese, *Wood Sci. Technol.*, 1975, **9**, 87.
28 K. Killham, *Soil Ecology*, Cambridge University Press, Cambridge, 1994.
29 R.L. Tate, *Soil Microbiology*, 2nd edition, John Wiley & Sons, Inc., New York, 2000.
30 D. Standing and K. Killham, in *Modern Soil Microbiology*, 2nd edition, ed. J. D. van Elsas, J. K. Jansson and J. T. Trevors, CRC Press. Taylor and Francis Group, Boca Raton, 2007, p. 1.
31 E.A. Bayer, H. Chanzy, R. Lamed, and Y. Shoham, *Curr. Opin. Struct. Biol.*, 1998, **8**, 548.
32 E.A. Bayer, B. Henrissat, and R. Lamed, in *Biomass Recalcitrance. Deconstructing the Plant Cell Wall for Bioenergy*, ed. M. E. Himmel, Blackwell Publishing, Oxford, 2008, p. 407.
33 H. Wei, Q. Xu, L.E. Taylor II, J.O. Baker, M.P. Tucker and S.Y. Ding, *Curr. Opin. Biotech.*, 2009, **20**, 330.

34 S.R. Decker, M. Siika-aho and L. Viikari, in *Biomass Recalcitrance. Deconstructing the Plant Cell Wall for Bioenergy*, ed. M. E. Himmel, Blackwell Publishing, Oxford, 2008, p. 352.

35 R. Benner, A.E. Maccubbin and R.E. Hodson, *Appl. Environ. Microbiol.*, 1984, **47**, 998.

36 J.J. Ko, Y. Shimizu, K. Ikeda, S.K. Kim, C.H. Park and S. Matsui, *Bioresource Technol.*, 2009, **100**, 1622.

37 S. Pareek, J.I. Azuma, S. Matsui and Y. Shimizu, *Water Sci. Technol.*, 2001, **44**, 351.

38 K.-E.L. Eriksson, R.A. Blanchette and P. Ander, 1990. *Microbial and Enzymatic Degradation of Wood and Wood Components,* Springer-Verlag, Berlin, 1990, ch. 1.4

39 R. Vicuña, *Enzym. Microb. Tech.*, 1998, **10**, 646.

40 L.Y. Young, and A.C. Frazer, *Geomicrobiol. J.*, 1987, **5**, 261.

41 J.G. Zeikus, A.L. Wellstein and T.K. Kirk, *FEMS Microbiol. Lett.*, 1982, **15**, 193.

42 W. Zimmermann, *J. Biotech.*, 1990, **13**, 119.

43 C.G. Björdal and T. Nilsson, in *Proceedings of the 8th ICOM Group on Wet Organic Archaeological Materials Conference, Stockholm 2001,* ed. P. Hoffmann, J.A. Spriggs, T. Grant, C. Cook, and A. Recht, The International Council of Museums, Committee for Conservation, Waterlogged Wood Working Group, Bremerhaven, 2002, p. 235.

44 Y.S. Kim and A.P. Singh, *Iawa J.*, 2000, **21**, 135.

45 A.P. Singh, T. Nilsson, and G.F. Daniel, *J. Inst. Wood Sci.*, 1990, **11**, 237.

46 R.A. Zabel and J.J. Morrell, *Wood Microbiology. Decay and Its Prevention.* San Diego, Academic Press, Inc., 1992, ch. 2.

47 D.M. Holt and E.B. Jones, *Appl. Environ. Microbiol.*, 1983, **46**, 722.

48 D.J. Huisman, M.R. Manders, E.I. Kretschmar, R.K.W.M. Klaassen and N. Lamersdorf, *Int. Biodeter. Biodegrad.*, 2008, **61**, 33.

49 Y.S. Kim, A.P. Singh and T. Nilsson, *Holzforschung*, 1996, **50**, 389.

50 R.K.W.M. Klaassen, *Int. Biodeter. Biodegrad.*, 2008, **61**, 45.

51 R.A. Blanchette and P. Hoffmann, in *Proceedings of the 5th ICOM Group on Wet Organic Archeological Materials, Portland/Maine 1993.* ed. P. Hoffmann, Ditzen Druck und Verlags-GmbH, Bremerhaven, 1994, p. 111.

52 A.P. Singh, Y.S. Kim, S.G. Wi, K.H. Lee and I.J. Kim, *Holzforschung*, 2003, **57**, 115.

53 O. Schmidt, U. Moreth and U. Schmitt, *Material und Organismen*, 1995, **29**, 289.

54 O. Schmidt, Y. Nagashima, W. Liese and U. Schmitt, *Holzforschung*, 1987, **41**, 137.

55 T. Nilsson, C. Björdal, and E. Fällman, *Int. Biodeter. Biodegrad.*, 2008, **61**, 17.

56 T. Nilsson and C. Björdal, *Int. Biodeter. Biodegrad.*, 2008, **61**, 3.

57 T. Nilsson and C. Björdal, *Int. Biodeter. Biodegrad.*, 2008, **61**, 11.

58 E.T. Landy, J.I. Mitchell, S. Hotchkiss and R.A. Eaton, *Int. Biodeter. Biodegrad.*, 2008, **61**, 106.

59 A.C. Helms, A.C. Martiny, J. Hofman-Bang, K. Ahring and M. Kilstrup, *Int. Biodeter. Biodegrad.*, 2004, **53**, 79.

60 A.C. Helms, *Bacterial Diversity in Waterlogged Archaeological Wood.* PhD thesis, The National Museum of Denmark / DTU Biosys, 2008.

61 P. Beguin and J.P. Aubert, *FEMS Microbiol. Rev.*, 1994, **13**, 25.

62 C.G. Björdal, G. Daniel, and T. Nilsson, *Int. Biodeter. Biodegrad.*, 2000, **45**, 15.

63 C.G. Björdal, *Waterlogged Archaeological Wood. Biodegradation and its Implications for conservation.* Doctoral thesis, Swedish University of Agricultural Science, 2000.

64 R.A. Eaton and M.D.C. Hale, *Wood. Decay, Pests and Protection*, 1st edn. Chapman & Hall, London, 1993

65 M.-L.E. Florian, in *ICOM committee for conservation. 6th triennial meeting, Ottawa, 21 - 25 September 1981. Preprints*, ed C. Pearson, The International Council of Museums, Paris 1981, p. 81.

66 J. Sen and R.K. Basak, *Geologiska föreningens i Stockholm förhandlingar*, 1957, **79**, 737.

67 P. Jensen and D.J. Gregory, *Archaeol. Sci.*, 2006, **33**, 551.

68 J.I. Hedges, in *Archaeological Wood*, ed. R.M. Rowell and R.J. Barbour, American Chemical Society, Washington D.C., 1990, p. 111.

69 A.P. Schniewind, in *Archaeological Wood*, ed. R.M. Rowell and R.J. Barbour, American Chemical Society, Washington D.C., 990, p. 87.

70 E.I. Kretschmar, J. Gelbrich, H. Militz and N. Lamersdorf, *Int. Biodeter. Biodegrad.*, 2008, **61**, 69.

71 C.G. Björdal, T. Nilsson and S. Bardage, *Holzforschung*, 2005, **59**, 178.

72 R.J. Barbour and L. Leney, in *The 6th International Biodeterioration Symposium, Washington D.C. 1984*. ed. S. Barry and D. R. Houghton, CAB International, Slough, United Kingdom, 1986, p. 189.

73 P. Hoffmann, A. Singh, Y.S. Kim, S.G. Wi, I.J. Kim and U. Schmitt, *Holzforschung*, 2004, **58**, 211.

74 D.M. Holt, *J. Inst. Wood Sci.*, 1983, **9**, 212.

75 G.F. Daniel and T. Nilsson, *Can. J. Microbiol.*, 1987, **33**, 943.

76 C.G. Björdal, *Int. Biodeter. Biodegrad.*, 2012, **70**, 126.

77 C. Capretti, N. Macchioni, B. Pizzo, G. Galotta, G. Giachi and D. Giampaola, *Archaeometry*, 2008, **50**, 855.

78 G. Daniel and T. Nilsson, in: *Forest Products Biotechnology*, ed. A. Bruce and J. W. Palfreyman, Taylor & Francis, London, 1998, p. 37.

79 J. Gelbrich, C. Mai and H. Militz, *Int. Biodeter. Biodegrad.*, 2008, **61**, 24.

80 G. Giachi and B. Pizzo, in *Proceedings of the 10th ICOM Group on Wet Organic Archaeological Materials Conference, Amsterdam 2007*, ed. K. Strætkvern and D. J. Huisman, Rijksdienst voor Archaeologie, Cultuurlandschap en Monumenten, Amersfoort, 2009, p. 21.

81 J.I. Hedges, G.L. Cowie, J.R. Ertel, R.J. Barbour and P.G. Hatcher, *Geochim. et Cosmochim. Acta*, 1985, **49**, 701.

82 D.W. Grattan and C. Mathias, *Somerset Levels Papers*, 1986, **12**, 7.

83 P. Hoffmann, in *Proceedings of the ICOM Waterlogged Wood Working Group Conference, Ottawa 1981*, ed. D. W. Grattan, The International Council of Museums, Committee for Conservation, Waterlogged Wood Working Group, Ottawa, 1982, p. 73.

84 K. Iiyama, N. Kasuya, L.T.B. Tuyet, J. Nakano and H. Sakaguchi, *Holzforschung*, 1988, **42**, 5.

85 Y.S. Kim, *Holzforschung*, 1990, **44**, 169.

86 D.R. Pan, D.S. Tai, C.K. Chen and D. Robert, *Holzforschung*, 1990, **44**, 7.

87 B. Pizzo, G. Giachi and L. Fiorentino, *Archaeometry*, 2010, **52**, 656.

88 K. Borgin, O. Faix, and W. Schweers, *Wood Sci. Technol.*, 1975, **9**, 207.

89 A.L. Kirillov and E.A. Mikolajchuk, in *9th Triennial Meeting for ICOM Committee for Conservation*, ed. Kirsten Grimstad, ICOM Committee for Conservation, Los Angeles, 1990, p. 239.

90 I.D. MacLeod and V.L. Richards in: *Proceedings of the 6th ICOM Group on Wet Organic Archaeological Materials Conference*, York 1996. ed. P. Hoffmann, T. Grant, J.A. Spriggs, T. Daley, The International Council of Museums Committee for

Conservation, Working Group on Wet Organic Archaeological Materials, Bremerhaven 1997.

91 H. Pavlikova, I. Sykorova, J. Cerny, E. Sebestova and V. Machovic, *Energy & Fuels*, 1993, **7**, 351.

92 A. Sandak, J. Sandak, M. Zborowska and W. Pradzynski, Presented at *COST Action IE0601 International Conference on Wooden Cultural Heritage: Evaluation of Deterioration and Management of Change. COST Action IE0601 - Wood Science for Conservation of Cultural Heritage*, Hamburg (DE), 7-10 October 2009, http://www.woodculther.org

93 M. Bardet, G. Gerbaud, M. Giffard, C. Doan, S. Hediger and L.L. Pape, *Prog. Nucl. Mag. Resonan. Spect.* 2009, **55**, 199.

94 M. Bardet, M.F. Foray, S. Maron, P. Goncalves and Q.-K. Trân, *Carbohyd. Polym.*, 2004, **57**, 419.

95 M.P. Colombini, M. Orlandi, F. Modugno, E.L. Tolppa, M. Sardelli, L. Zoia and C. Crestini, *Microchem. J.*, 2007, **85**, 164.

96 M.P. Colombini, J.J. Lucejko, F. Modugno, M. Orlandi, E.L. Tolppa and L. Zoia, *Talanta*, 2009, **80**, 61.

97 P.G. Hatcher, I.A. Breger and W.L. Earl, *Organic Geochem.*, 1981, **3**, 49.

98 A. Salanti, L. Zoia, E.L. Tolppa and M. Orlandi, in *Italic 5 - Science & Technology of Biomasses: Advances and Challenges. From Forest and Agricultural Biomasses to High Added Value Products: Processes and Materials*, September 1-4, 2009, Villa Monasterio, Varenna (Como), Italy, ed. M. Orlandi and C. Crestini, Éxòrma, 2009, p. 63.

99 A. Salanti, L. Zoia, E.L. Tolppa, G. Giachi and M. Orlandi, *Microchem. J.*, 2010, **95**, 345.

100 M.A. Wilson, I.M. Godfrey, J.V. Hanna, R.A. Quezada and K.S. Finnie, K.S., *Organic Geochemistry*, 1993, **20**, 599.

101 J.J. Lucejko, *Waterlogged Archaeological Wood: Chemical Study of Wood Degradation and Evaluation of Consolidation Treatments.* PhD thesis, Pisa University, 2010.

102 V. Richards and N. West, in *Proceedings of the 8th ICOM Group on Wet Organic Archaeological Materials Conference, Stockholm 2001*, ed. P. Hoffmann, J.A. Spriggs, T. Grant, C. Cook and A. Recht, The International Council of Museums, Committee for Conservation, Waterlogged Wood Working Group Bremerhaven: 2002, p. 259.

103 C. Saiz-Jimenez, J.J. Boon, J.I. Hedges, J.K.C. Hessels and J.W. De Leeuw, *J. Analytic. Appl. Pyrolysis*, 1987, **11**, 437.

104 P.F. van Bergen, I. Poole, T.M.A. Ogilvie, C. Caple and R.P. Evershed, *Rapid Commun. Mass Spectrom.*, 2000, **14**, 71.

105 M.J. Pelletier and C.C. Pelletier, in *Raman, Infrared, and Near-Infrared Chemical Imaging*, ed. S. Šašic and Y. Ozaki, John Wiley & Sons, Inc., Hoboken, NJ, USA, 2010, p. 1.

106 N. Gierlinger and M. Schwanninger, *Spectroscopy: an International Journal*, 2007, **21**, 69.

107 I. Irbe, G. Noldt, G. Koch, I. Andersone and B. Andersons, *Holzforschung*, 2006, **60**, 601.

108 K. Cufar, J. Gricar, M. Zupancic, G. Koch and U. Schmitt, *Iawa J.*, 2008, **29**, 55.

109 N. Pedersen (unpublished data)

110 A.P. Singh and Y.S. Kim, Paper prepared for the 28th Annual Meeting, The

International Research Group on Wood Preservation. IRG/WP 97-10217, Whistle, Canada 25-30 May 1997. International Research Group on Wood Preservation, p. 1.

111 M.P. Colombini, J.J. Lucejko, F. Modugno and E. Ribechini, in: *Proceedings of the 10th ICOM Group on Wet Organic Archaeological Materials Conference, Amsterdam 2007*, ed. K. Strætkvern and D. J. Huisman, Rijksdienst voor Archaeologie, Cultuurlandschap en Monumenten, Amersfoort, 2009, p. 35.

112 J. Gelbrich, M. Carsten and H. Militz, Presented at *COST Action IE0601 International Conference on Wooden Cultural Heritage: Evaluation of Deterioration and Management of Change. COST Action IE0601 - Wood Science for Conservation of Cultural Heritage*, Hamburg (DE), 7-10 October 2009, http://www.woodculther.org

113 F. Modugno, J.J. Lucejko, E. Ribechini, M.P. Colombini and J.C. Del Rio, in *Italic 5 - Science & Technology of Biomasses: Advances and Challenges. From Forest and Agricultural Biomasses to High Added Value Products: Processes and Materials*, September 1-4, 2009, Villa Monasterio, Varenna (Como), Italy, ed. M. Orlandi and C. Crestini, Éxòrma, 2009, p. 27.

114 A. Sandak, J. Sandak, M. Zborowska and W. Pradzynski, *J. Archaeol. Sci.*, 2010, **37**, 2093.

115 A. Thygesen, J. Oddershede, H. Lilholt, A.B. Thomsen and K. Stahl, *Cellulose*, 2005, **12**, 563.

116 J.J. Lucejko, F. Modugno, E. Ribechini and J.C. del Rio, *Analytic. Chim. Acta*, 2009, **654**, 26.

117 D. Fengel, *Wood Sci. Technol.*, 1991, **25**, 153.

118 G. Giachi, F. Bettazzi, S. Chimichi and G. Staccioli, *J. Cultural Heritage*, 2003, **4**, 75.

119 Y. Fors, T. Nilsson, E.D. Risberg, M. Sandström and P. Torssander, *Int. Biodeter. Biodegrad.*, 2008, **62**, 336.

120 B.A. Jordan, *Int. Biodeter. Biodegrad.*, 2001, **47**, 47.

121 R.K.W.M. Klaassen, *Int. Biodeter. Biodegrad.*, 2008, **61**, 61.

122 C.G. Björdal, C.G. and T. Nilsson, *J. Archaeol. Sci.*, 2008, **35**, 862.

123 R.K.W.M. Klaassen, in *Proceedings of the 10th ICOM Group on Wet Organic Archaeological Materials Conference, Amsterdam 2007*. ed. K. Strætkvern and D. J. Huisman, Rijksdienst voor Archaeologie, Cultuurlandschap en Monumenten, Amersfoort, 2009, p. 69.

124 T. Nilsson, in: *The 7th ICOM Group on Wet Organic Archaeological Materials Conference, Grenoble 1998*, ed. C. Bonnot-Diconne, X. Hiiron, Q.K. Tran, P. Hoffman, ARC-Nucléart CEA/Grenoble, 1999, p. 65.

125 D.J. Huisman, E.I. Kretschmar and N. Lamersdorf, *Int. Biodeter. Biodegrad.*, 2008, **61**, 117.

126 I.N. Godfrey, T. Bergstrand, C.G. Bjordal, T. Nilsson, C. Bohm, E. Christensen, D. Gregory, E.E. Peacock, V. Richards and I.D. MacLeod, in *Proceedings of the 10th ICOM Group on Wet Organic Archaeological Materials, Amsterdam 2007*, ed. K. Strætkvern and D. J. Huisman, Rijkdienst voor Archaeologie, Cultuurlandschap en Monumenten, Amersfoort, 2009, p. 169.

127 D. Gregory and M. Manders, In: *Wreck Protect. Decay and Protection of Archaeological Wooden Shipwrecks*, ed. C. G. Björdal and D. Gregory, Archaeopress Ltd., Oxford, 2011, p. 107.

Subject Index